FREE Study Skills DVD Offer

Dear Customer,

Thank you for your purchase from Mometrix! We consider it an honor and privilege that you have purchased our product and want to ensure your satisfaction.

As a way of showing our appreciation and to help us better serve you, we have developed a Study Skills DVD that we would like to give you for <u>FREE</u>. **This DVD covers our "best practices" for studying for your exam, from using our study materials to preparing for the day of the test.**

All that we ask is that you email us your feedback that would describe your experience so far with our product. Good, bad or indifferent, we want to know what you think!

To get your **FREE Study Skills DVD**, email <u>freedvd@mometrix.com</u> with "FREE STUDY SKILLS DVD" in the subject line and the following information in the body of the email:

    a. The name of the product you purchased.

    b. Your product rating on a scale of 1-5, with 5 being the highest rating.

    c. Your feedback. It can be long, short, or anything in-between, just your impressions and experience so far with our product. Good feedback might include how our study material met your needs and will highlight features of the product that you found helpful.

    d. Your full name and shipping address where you would like us to send your free DVD.

If you have any questions or concerns, please don't hesitate to contact me directly.

Thanks again!

Sincerely,

Jay Willis
Vice President
jay.willis@mometrix.com
1-800-673-8175

Praxis II

Mathematics: Content Knowledge (5161) Exam

SECRETS

Study Guide
Your Key to Exam Success

Praxis II Test Review for the
Praxis II: Subject Assessments

Published by
Mometrix Test Preparation
Praxis II Exam Secrets Test Prep Team

Written and edited by the Praxis II Exam Secrets Test Prep Staff

Printed in the United States of America

This paper meets the requirements of ANSI/NISO Z39.48-1992 (Permanence of Paper).

Mometrix offers volume discount pricing to institutions. For more information or a price quote, please contact our sales department at sales@mometrix.com or 888-248-1219.

Praxis II® is a registered trademark of Educational Testing Services ® (ETS®), which was not involved in the production of, and does not endorse, this product.

ISBN 13: 978-1-63094-510-7
ISBN 10: 1-63094-510-2

Dear Future Exam Success Story:

Congratulations on your purchase of our study guide. Our goal in writing our study guide was to cover the content on the test, as well as provide insight into typical test taking mistakes and how to overcome them.

Standardized tests are a key component of being successful, which only increases the importance of doing well in the high-pressure high-stakes environment of test day. How well you do on this test will have a significant impact on your future, and we have the research and practical advice to help you execute on test day.

The product you're reading now is designed to exploit weaknesses in the test itself, and help you avoid the most common errors test takers frequently make.

How to use this study guide

We don't want to waste your time. Our study guide is fast-paced and fluff-free. We suggest going through it a number of times, as repetition is an important part of learning new information and concepts.

First, read through the study guide completely to get a feel for the content and organization. Read the general success strategies first, and then proceed to the content sections. Each tip has been carefully selected for its effectiveness.

Second, read through the study guide again, and take notes in the margins and highlight those sections where you may have a particular weakness.

Finally, bring the manual with you on test day and study it before the exam begins.

Your success is our success

We would be delighted to hear about your success. Send us an email and tell us your story. Thanks for your business and we wish you continued success.

Sincerely,

Mometrix Test Preparation Team

Need more help? Check out our flashcards at: http://MometrixFlashcards.com/PraxisII

TABLE OF CONTENTS

Top 20 Test Taking Tips

1. Carefully follow all the test registration procedures
2. Know the test directions, duration, topics, question types, how many questions
3. Setup a flexible study schedule at least 3-4 weeks before test day
4. Study during the time of day you are most alert, relaxed, and stress free
5. Maximize your learning style; visual learner use visual study aids, auditory learner use auditory study aids
6. Focus on your weakest knowledge base
7. Find a study partner to review with and help clarify questions
8. Practice, practice, practice
9. Get a good night's sleep; don't try to cram the night before the test
10. Eat a well balanced meal
11. Know the exact physical location of the testing site; drive the route to the site prior to test day
12. Bring a set of ear plugs; the testing center could be noisy
13. Wear comfortable, loose fitting, layered clothing to the testing center; prepare for it to be either cold or hot during the test
14. Bring at least 2 current forms of ID to the testing center
15. Arrive to the test early; be prepared to wait and be patient
16. Eliminate the obviously wrong answer choices, then guess the first remaining choice
17. Pace yourself; don't rush, but keep working and move on if you get stuck
18. Maintain a positive attitude even if the test is going poorly
19. Keep your first answer unless you are positive it is wrong
20. Check your work, don't make a careless mistake

Numbers and Quantity / Algebra

Numbers and their Classifications

Numbers are the basic building blocks of mathematics. Specific features of numbers are identified by the following terms:

Integers – The set of whole positive and negative numbers, including zero. Integers do not include fractions $\left(\frac{1}{3}\right)$, decimals (0.56), or mixed numbers $\left(7\frac{3}{4}\right)$.

Prime number – A whole number greater than 1 that has only two factors, itself and 1; that is, a number that can be divided evenly only by 1 and itself.

Composite number – A whole number greater than 1 that has more than two different factors; in other words, any whole number that is not a prime number. For example: The composite number 8 has the factors of 1, 2, 4, and 8.

Even number – Any integer that can be divided by 2 without leaving a remainder. For example: 2, 4, 6, 8, and so on.

Odd number – Any integer that cannot be divided evenly by 2. For example: 3, 5, 7, 9, and so on.

Decimal number – a number that uses a decimal point to show the part of the number that is less than one. Example: 1.234.

Decimal point – a symbol used to separate the ones place from the tenths place in decimals or dollars from cents in currency.

Decimal place – the position of a number to the right of the decimal point. In the decimal 0.123, the 1 is in the first place to the right of the decimal point, indicating tenths; the 2 is in the second place, indicating hundredths; and the 3 is in the third place, indicating thousandths.

The decimal, or base 10, system is a number system that uses ten different digits (0, 1, 2, 3, 4, 5, 6, 7, 8, 9). An example of a number system that uses something other than ten digits is the binary, or base 2, number system, used by computers, which uses only the numbers 0 and 1. It is thought that the decimal system originated because people had only their 10 fingers for counting.

Rational, irrational, and real numbers can be described as follows:

Rational numbers include all integers, decimals, and fractions. Any terminating or repeating decimal number is a rational number.

Irrational numbers cannot be written as fractions or decimals because the number of decimal places is infinite and there is no recurring pattern of digits within the number. For example, pi (π) begins with 3.141592 and continues without terminating or repeating, so pi is an irrational number.

Real numbers are the set of all rational and irrational numbers.

Operations

There are four basic mathematical operations:

Addition increases the value of one quantity by the value of another quantity. Example: $2 + 4 = 6; 8 + 9 = 17$. The result is called the sum. With addition, the order does not matter. $4 + 2 = 2 + 4$.

Subtraction is the opposite operation to addition; it decreases the value of one quantity by the value of another quantity. Example: $6 - 4 = 2; 17 - 8 = 9$. The result is called the difference. Note that with subtraction, the order does matter. $6 - 4 \neq 4 - 6$.

Multiplication can be thought of as repeated addition. One number tells how many times to add the other number to itself. Example: 3×2 (three times two) $= 2 + 2 + 2 = 6$. With multiplication, the order does not matter. $2 \times 3 = 3 \times 2$ or $3 + 3 = 2 + 2 + 2$.

Division is the opposite operation to multiplication; one number tells us how many parts to divide the other number into. Example: $20 \div 4 = 5$; if 20 is split into 4 equal parts, each part is 5. With division, the order of the numbers does matter. $20 \div 4 \neq 4 \div 20$.

An exponent is a superscript number placed next to another number at the top right. It indicates how many times the base number is to be multiplied by itself. Exponents provide a shorthand way to write what would be a longer mathematical expression. Example: $a^2 = a \times a$; $2^4 = 2 \times 2 \times 2 \times 2$. A number with an exponent of 2 is said to be "squared," while a number with an exponent of 3 is said to be "cubed." The value of a number raised to an exponent is called its power. So, 8^4 is read as "8 to the 4th power," or "8 raised to the power of 4." A negative exponent is the same as the reciprocal of a positive exponent. Example: $a^{-2} = \frac{1}{a^2}$.

> **Review Video: <u>Exponents</u>**
> *Visit **mometrix.com/academy** and enter **Code: 600998**

Order of Operations is a set of rules that dictates the order in which we must perform each operation in an expression so that we will evaluate at accurately. If we have an expression that includes multiple different operations, Order of Operations tells us which operations to do first. The most common mnemonic for Order of Operations is PEMDAS, or "Please Excuse My Dear Aunt Sally." PEMDAS stands for Parentheses, Exponents, Multiplication, Division, Addition, Subtraction. It is important to understand that multiplication and division have equal precedence, as do addition and subtraction, so those pairs of operations are simply worked from left to right in order.

> **Review Video: <u>Order of Operations</u>**
> *Visit **mometrix.com/academy** and enter **Code: 259675**

Example: Evaluate the expression $5 + 20 \div 4 \times (2 + 3)^2 - 6$ using the correct order of operations.
P: Perform the operations inside the parentheses, $(2 + 3) = 5$.
E: Simplify the exponents, $(5)^2 = 25$.
The equation now looks like this: $5 + 20 \div 4 \times 25 - 6$.
MD: Perform multiplication and division from left to right, $20 \div 4 = 5$; then $5 \times 25 = 125$.
The equation now looks like this: $5 + 125 - 6$.
AS: Perform addition and subtraction from left to right, $5 + 125 = 130$; then $130 - 6 = 124$.

The laws of exponents are as follows:
1) Any number to the power of 1 is equal to itself: $a^1 = a$.
2) The number 1 raised to any power is equal to 1: $1^n = 1$.
3) Any number raised to the power of 0 is equal to 1: $a^0 = 1$.
4) Add exponents to multiply powers of the same base number: $a^n \times a^m = a^{n+m}$.
5) Subtract exponents to divide powers of the same number; that is $a^n \div a^m = a^{n-m}$.
6) Multiply exponents to raise a power to a power: $(a^n)^m = a^{n \times m}$.
7) If multiplied or divided numbers inside parentheses are collectively raised to a power, this is the same as each individual term being raised to that power: $(a \times b)^n = a^n \times b^n$; $(a \div b)^n = a^n \div b^n$.
Note: Exponents do not have to be integers. Fractional or decimal exponents follow all the rules above as well. Example: $5^{\frac{1}{4}} \times 5^{\frac{3}{4}} = 5^{\frac{1}{4} + \frac{3}{4}} = 5^1 = 5$.

A root, such as a square root, is another way of writing a fractional exponent. Instead of using a superscript, roots use the radical symbol ($\sqrt{\ }$) to indicate the operation. A radical will have a number underneath the bar, and may sometimes have a number in the upper left: $\sqrt[n]{a}$, read as "the nth root of a." The relationship between radical notation and exponent notation can be described by this equation: $\sqrt[n]{a} = a^{\frac{1}{n}}$. The two special cases of $n = 2$ and $n = 3$ are called square roots and cube roots. If there is no number to the upper left, it is understood to be a square root ($n = 2$). Nearly all of the roots you encounter will be square roots. A square root is the same as a number raised to the one-half power. When we say that a is the square root of b ($a = \sqrt{b}$), we mean that a multiplied by itself equals b: ($a \times a = b$).

> ➤ **Review Video: <u>Square Root and Perfect Square</u>**
> *Visit **mometrix.com/academy** and enter **Code**: **648063***

A perfect square is a number that has an integer for its square root. There are 10 perfect squares from 1 to 100: 1, 4, 9, 16, 25, 36, 49, 64, 81, 100 (the squares of integers 1 through 10). Parentheses are used to designate which operations should be done first when there are multiple operations. Example: 4 – (2 + 1) = 1; the parentheses tell us that we must add 2 and 1, and then subtract the sum from 4, rather than subtracting 2 from 4 and then adding 1 (this would give us an answer of 3).

Scientific notation is a way of writing large numbers in a shorter form. The form $a \times 10^n$ is used in scientific notation, where a is greater than or equal to 1, but less than 10, and n is the number of places the decimal must move to get from the original number to a. Example: The number 230,400,000 is cumbersome to write. To write the value in scientific notation, place a decimal point between the first and second numbers, and include all digits through the last non-zero digit ($a = 2.304$). To find the appropriate power of 10, count the number of places the decimal point had to move ($n = 8$). The number is positive if the decimal moved to the left, and negative if it moved to the right. We can then write 230,400,000 as 2.304×10^8. If we look instead at the number 0.00002304, we have the same value for a, but this time the decimal moved 5 places to the right ($n = -5$). Thus, 0.00002304 can be written as 2.304×10^{-5}. Using this notation makes it simple to compare very large or very small numbers. By comparing exponents, it is easy to see that 3.28×10^4 is smaller than 1.51×10^5, because 4 is less than 5.

> ➤ **Review Video: <u>Scientific Notation</u>**
> *Visit **mometrix.com/academy** and enter **Code**: **976454***

Positive and Negative Numbers

A precursor to working with negative numbers is understanding what absolute values are. A number's *Absolute Value* is simply the distance away from zero a number is on the number line. The absolute value of a number is always positive and is written $|x|$.

When adding signed numbers, if the signs are the same, simply add the absolute values of the addends and apply the original sign to the sum. For example, (+4) + (+8) = +12 and (−4) + (−8) = −12. When the original signs are different, take the absolute values of the addends and subtract the smaller value from the larger value, then apply the original sign of the larger value to the difference. For instance, (+4) + (−8) = −4 and (−4) + (+8) = +4.

For subtracting signed numbers, change the sign of the number after the minus symbol and then follow the same rules used for addition. For example, $(+4)-(+8) = (+4) + (-8) = -4$.

If the signs are the same the product is positive when multiplying signed numbers. For example, $(+4) \times (+8) = +32$ and $(-4) \times (-8) = +32$. If the signs are opposite, the product is negative. For example, $(+4) \times (-8) = -32$ and $(-4) \times (+8) = -32$. When more than two factors are multiplied together, the sign of the product is determined by how many negative factors are present. If there are an odd number of negative factors then the product is negative, whereas an even number of negative factors indicates a positive product. For instance, $(+4) \times (-8) \times (-2) = +64$ and $(-4) \times (-8) \times (-2) = -64$.

The rules for dividing signed numbers are similar to multiplying signed numbers. If the dividend and divisor have the same sign, the quotient is positive. If the dividend and divisor have opposite signs, the quotient is negative. For example, $(-4) \div (+8) = -0.5$.

Factors and Multiples

Factors are numbers that are multiplied together to obtain a product. For example, in the equation $2 \times 3 = 6$, the numbers 2 and 3 are factors. A prime number has only two factors (1 and itself), but other numbers can have many factors.

A common factor is a number that divides exactly into two or more other numbers. For example, the factors of 12 are 1, 2, 3, 4, 6, and 12, while the factors of 15 are 1, 3, 5, and 15. The common factors of 12 and 15 are 1 and 3.

A prime factor is also a prime number. Therefore, the prime factors of 12 are 2 and 3. For 15, the prime factors are 3 and 5.

The greatest common factor (GCF) is the largest number that is a factor of two or more numbers. For example, the factors of 15 are 1, 3, 5, and 15; the factors of 35 are 1, 5, 7, and 35. Therefore, the greatest common factor of 15 and 35 is 5.

> ➤ **Review Video: <u>Greatest Common Factor (GCF)</u>**
> *Visit **mometrix.com/academy** and enter **Code**: **838699***

The least common multiple (LCM) is the smallest number that is a multiple of two or more numbers. For example, the multiples of 3 include 3, 6, 9, 12, 15, etc.; the multiples of 5 include 5, 10, 15, 20, etc. Therefore, the least common multiple of 3 and 5 is 15.

Fractions, Percentages, and Related Concepts

A fraction is a number that is expressed as one integer written above another integer, with a dividing line between them $\left(\frac{x}{y}\right)$. It represents the quotient of the two numbers "x divided by y." It can also be thought of as x out of y equal parts.

The top number of a fraction is called the numerator, and it represents the number of parts under consideration. The 1 in $\frac{1}{4}$ means that 1 part out of the whole is being considered in the calculation. The bottom number of a fraction is called the denominator, and it represents the total number of equal parts. The 4 in $\frac{1}{4}$ means that the whole consists of 4 equal parts. A fraction cannot have a denominator of zero; this is referred to as "undefined."

> **Review Video:** <u>Fractions</u>
> Visit **mometrix.com/academy** and enter **Code: 262335**

Fractions can be manipulated, without changing the value of the fraction, by multiplying or dividing (but not adding or subtracting) both the numerator and denominator by the same number. If you divide both numbers by a common factor, you are reducing or simplifying the fraction. Two fractions that have the same value, but are expressed differently are known as equivalent fractions. For example, $\frac{2}{10}, \frac{3}{15}, \frac{4}{20}$, and $\frac{5}{25}$ are all equivalent fractions. They can also all be reduced or simplified to $\frac{1}{5}$.

When two fractions are manipulated so that they have the same denominator, this is known as finding a common denominator. The number chosen to be that common denominator should be the least common multiple of the two original denominators. Example: $\frac{3}{4}$ and $\frac{5}{6}$; the least common multiple of 4 and 6 is 12. Manipulating to achieve the common denominator: $\frac{3}{4} = \frac{9}{12}; \frac{5}{6} = \frac{10}{12}$.

If two fractions have a common denominator, they can be added or subtracted simply by adding or subtracting the two numerators and retaining the same denominator. Example: $\frac{1}{2} + \frac{1}{4} = \frac{2}{4} + \frac{1}{4} = \frac{3}{4}$. If the two fractions do not already have the same denominator, one or both of them must be manipulated to achieve a common denominator before they can be added or subtracted.

Two fractions can be multiplied by multiplying the two numerators to find the new numerator and the two denominators to find the new denominator. Example: $\frac{1}{3} \times \frac{2}{3} = \frac{1 \times 2}{3 \times 3} = \frac{2}{9}$.

> **Review Video:** <u>Multiplying Fractions</u>
> Visit **mometrix.com/academy** and enter **Code: 638849**

Two fractions can be divided flipping the numerator and denominator of the second fraction and then proceeding as though it were a multiplication. Example: $\frac{2}{3} \div \frac{3}{4} = \frac{2}{3} \times \frac{4}{3} = \frac{8}{9}$.

> **Review Video:** <u>Dividing Fractions</u>
> Visit **mometrix.com/academy** and enter **Code: 300874**

A fraction whose denominator is greater than its numerator is known as a proper fraction, while a fraction whose numerator is greater than its denominator is known as an improper fraction. Proper fractions have values less than one and improper fractions have values greater than one.

A mixed number is a number that contains both an integer and a fraction. Any improper fraction can be rewritten as a mixed number. Example: $\frac{8}{3} = \frac{6}{3} + \frac{2}{3} = 2 + \frac{2}{3} = 2\frac{2}{3}$. Similarly, any mixed number can be rewritten as an improper fraction. Example: $1\frac{3}{5} = 1 + \frac{3}{5} = \frac{5}{5} + \frac{3}{5} = \frac{8}{5}$.

> **Review Video:** <u>Improper Fractions and Mixed Numbers</u>
> Visit **mometrix.com/academy** and enter **Code: 731507**

Percentages can be thought of as fractions that are based on a whole of 100; that is, one whole is equal to 100%. The word percent means "per hundred." Fractions can be expressed as percents by finding equivalent fractions with a denomination of 100. Example: $\frac{7}{10} = \frac{70}{100} = 70\%$; $\frac{1}{4} = \frac{25}{100} = 25\%$.

To express a percentage as a fraction, divide the percentage number by 100 and reduce the fraction to its simplest possible terms. Example: $60\% = \frac{60}{100} = \frac{3}{5}$; $96\% = \frac{96}{100} = \frac{24}{25}$.

Converting decimals to percentages and percentages to decimals is as simple as moving the decimal point. To convert from a decimal to a percent, move the decimal point two places to the right. To convert from a percent to a decimal, move it two places to the left. Example: 0.23 = 23%; 5.34 = 534%; 0.007 = 0.7%; 700% = 7.00; 86% = 0.86; 0.15% = 0.0015.

It may be helpful to remember that the percentage number will always be larger than the equivalent decimal number.

A percentage problem can be presented three main ways: (1) Find what percentage of some number another number is. Example: What percentage of 40 is 8? (2) Find what number is some percentage of a given number. Example: What number is 20% of 40? (3) Find what number another number is a given percentage of. Example: What number is 8 20% of? The three components in all of these cases are the same: a whole (W), a part (P), and a percentage (%). These are related by the equation: $P = W \times \%$. This is the form of the equation you would use to solve problems of type (2). To solve types (1) and (3), you would use these two forms: $\% = \frac{P}{W}$ and $W = \frac{P}{\%}$.

> ➤ **Review Video: <u>Percentages</u>**
> *Visit **mometrix.com/academy** and enter **Code**: 141911*

The thing that frequently makes percentage problems difficult is that they are most often also word problems, so a large part of solving them is figuring out which quantities are what. Example: In a school cafeteria, 7 students choose pizza, 9 choose hamburgers, and 4 choose tacos. Find the percentage that chooses tacos. To find the whole, you must first add all of the parts: 7 + 9 + 4 = 20. The percentage can then be found by dividing the part by the whole $\left(\% = \frac{P}{W}\right)$: $\frac{4}{20} = \frac{20}{100} = 20\%$.

A ratio is a comparison of two quantities in a particular order. Example: If there are 14 computers in a lab, and the class has 20 students, there is a student to computer ratio of 20 to 14, commonly written as 20:14. Ratios are normally reduced to their smallest whole number representation, so 20:14 would be reduced to 10:7 by dividing both sides by 2.

A proportion is a relationship between two quantities that dictates how one changes when the other changes. A direct proportion describes a relationship in which a quantity increases by a set amount for every increase in the other quantity, or decreases by that same amount for every decrease in the other quantity. Example: Assuming a constant driving speed, the time required for a car trip increases as the distance of the trip increases. The distance to be traveled and the time required to travel are directly proportional.

Inverse proportion is a relationship in which an increase in one quantity is accompanied by a decrease in the other, or vice versa. Example: the time required for a car trip decreases as the speed increases, and increases as the speed decreases, so the time required is inversely proportional to the speed of the car.

Equations and Graphing

When algebraic functions and equations are shown graphically, they are usually shown on a *Cartesian Coordinate Plane*. The Cartesian coordinate plane consists of two number lines placed perpendicular to each other, and intersecting at the zero point, also known as the origin. The horizontal number line is known as the *x*-axis, with positive values to the right of the origin, and negative values to the left of the origin. The vertical number line is known as the *y*-axis, with positive values above the origin, and negative values below the origin. Any point on the plane can be identified by an ordered pair in the form (*x,y*), called coordinates. The *x*-value of the coordinate is called the abscissa, and the *y*-value of the coordinate is called the ordinate. The two number lines divide the plane into four quadrants: I, II, III, and IV.

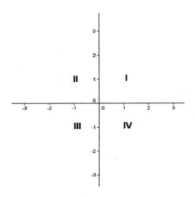

Before learning the different forms equations can be written in, it is important to understand some terminology. A ratio of the change in the vertical distance to the change in horizontal distance is called the *Slope*. On a graph with two points, (x_1, y_1) and (x_2, y_2), the slope is represented by the formula $s = \frac{y_2 - y_1}{x_2 - x_1}$; $x_1 \neq x_2$. If the value of the slope is positive, the line slopes upward from left to right. If the value of the slope is negative, the line slopes downward from left to right. If the *y*-coordinates are the same for both points, the slope is 0 and the line is a *Horizontal Line*. If the *x*-coordinates are the same for both points, there is no slope and the line is a *Vertical Line*. Two or more lines that have equal slopes are *Parallel Lines*. *Perpendicular Lines* have slopes that are negative reciprocals of each other, such as $\frac{a}{b}$ and $\frac{-b}{a}$.

Equations are made up of monomials and polynomials. A *Monomial* is a single variable or product of constants and variables, such as x, $2x$, or $\frac{2}{x}$. There will never be addition or subtraction symbols in a monomial. Like monomials have like variables, but they may have different coefficients. *Polynomials* are algebraic expressions which use addition and subtraction to combine two or more monomials. Two terms make a binomial; three terms make a trinomial; etc.. The *Degree of a Monomial* is the sum of the exponents of the variables. The *Degree of a Polynomial* is the highest degree of any individual term.

As mentioned previously, equations can be written many ways. Below is a list of the many forms equations can take.

- *Standard Form*: $Ax + By = C$; the slope is $\frac{-A}{B}$ and the y-intercept is $\frac{C}{B}$
- *Slope Intercept Form*: $y = mx + b$, where m is the slope and b is the y-intercept
- *Point-Slope Form*: $y - y_1 = m(x - x_1)$, where m is the slope and (x_1, y_1) is a point on the line
- *Two-Point Form*: $\frac{y-y_1}{x-x_1} = \frac{y_2-y_1}{x_2-x_1}$, where (x_1, y_1) and (x_2, y_2) are two points on the given line
- *Intercept Form*: $\frac{x}{x_1} + \frac{y}{y_1} = 1$, where $(x_1, 0)$ is the point at which a line intersects the x-axis, and $(0, y_1)$ is the point at which the same line intersects the y-axis

> ➤ **Review Video: <u>Slope Intercept and Point-Slope Forms</u>**
> *Visit **mometrix.com/academy** and enter **Code: 113216***

Equations can also be written as $ax + b = 0$, where $a \neq 0$. These are referred to as *One Variable Linear Equations*. A solution to such an equation is called a *Root*. In the case where we have the equation $5x + 10 = 0$, if we solve for x we get a solution of $x = -2$. In other words, the root of the equation is -2. This is found by first subtracting 10 from both sides, which gives $5x = -10$. Next, simply divide both sides by the coefficient of the variable, in this case 5, to get $x = -2$. This can be checked by plugging -2 back into the original equation $(5)(-2) + 10 = -10 + 10 = 0$.

The *Solution Set* is the set of all solutions of an equation. In our example, the solution set would simply be -2. If there were more solutions (there usually are in multivariable equations) then they would also be included in the solution set. When an equation has no true solutions, this is referred to as an *Empty Set*. Equations with identical solution sets are *Equivalent Equations*. An *Identity* is a term whose value or determinant is equal to 1.

Other Important Concepts

Commonly in algebra and other upper-level fields of math you find yourself working with mathematical expressions that do not equal each other. The statement comparing such expressions with symbols such as < (less than) or > (greater than) is called an *Inequality*. An example of an inequality is $7x > 5$. To solve for x, simply divide both sides by 7 and the solution is shown to be $x > \frac{5}{7}$. Graphs of the solution set of inequalities are represented on a number line. Open circles are used to show that an expression approaches a number but is never quite equal to that number.

> ➤ **Review Video: <u>Inequalities</u>**
> *Visit **mometrix.com/academy** and enter **Code: 451494***

Conditional Inequalities are those with certain values for the variable that will make the condition true and other values for the variable where the condition will be false. *Absolute Inequalities* can have any real number as the value for the variable to make the condition true, while there is no real number value for the variable that will make the condition false. Solving inequalities is done by following the same rules as for solving equations with the exception that when multiplying or dividing by a negative number the direction of the inequality sign must be flipped or reversed. *Double Inequalities* are situations where two inequality statements apply to the same variable expression. An example of this is $-c < ax + b < c$.

A *Weighted Mean*, or weighted average, is a mean that uses "weighted" values. The formula is weighted mean $= \frac{w_1 x_1 + w_2 x_2 + w_3 x_3 \dots + w_n x_n}{w_1 + w_2 + w_3 + \dots + w_n}$. Weighted values, such as $w_1, w_2, w_3, \dots w_n$ are assigned to each member of the set $x_1, x_2, x_3, \dots x_n$. If calculating weighted mean, make sure a weight value for each member of the set is used.

Calculations Using Points

Sometimes you need to perform calculations using only points on a graph as input data. Using points, you can determine what the midpoint and distance are. If you know the equation for a line you can calculate the distance between the line and the point.

To find the *Midpoint* of two points (x_1, y_1) and (x_2, y_2), average the x-coordinates to get the x-coordinate of the midpoint, and average the y-coordinates to get the y-coordinate of the midpoint. The formula is Midpoint $= \left(\frac{x_1 + x_2}{2}, \frac{y_1 + y_2}{2} \right)$.

The *Distance* between two points is the same as the length of the hypotenuse of a right triangle with the two given points as endpoints, and the two sides of the right triangle parallel to the x-axis and y-axis, respectively. The length of the segment parallel to the x-axis is the difference between the x-coordinates of the two points. The length of the segment parallel to the y-axis is the difference between the y-coordinates of the two points. Use the Pythagorean Theorem $a^2 + b^2 = c^2$ or $c = \sqrt{a^2 + b^2}$ to find the distance. The formula is Distance $= \sqrt{(x_2 - x_1)^2 + (y_2 - y_1)^2}$.

When a line is in the format $Ax + By + C = 0$, where A, B, and C are coefficients, you can use a point (x_1, y_1) not on the line and apply the formula $d = \frac{|Ax_1 + By_1 + C|}{\sqrt{A^2 + B^2}}$ to find the distance between the line and the point (x_1, y_1).

Systems of Equations

Systems of Equations are a set of simultaneous equations that all use the same variables. A solution to a system of equations must be true for each equation in the system. *Consistent Systems* are those with at least one solution. *Inconsistent Systems* are systems of equations that have no solution.

> ➤ **Review Video: <u>Systems of Equations</u>**
> Visit ***mometrix.com/academy*** *and enter* **Code**: **658153**

To solve a system of linear equations by *substitution*, start with the easier equation and solve for one of the variables. Express this variable in terms of the other variable. Substitute this expression in the other equation, and solve for the other variable. The solution should be expressed in the form (x, y). Substitute the values into both of the original equations to check your answer. Consider the following problem.

Solve the system using substitution:
$$x + 6y = 15$$
$$3x - 12y = 18$$

Solve the first equation for x:
$$x = 15 - 6y$$

Substitute this value in place of x in the second equation, and solve for y:
$$3(15 - 6y) - 12y = 18$$
$$45 - 18y - 12y = 18$$
$$30y = 27$$
$$y = \frac{27}{30} = \frac{9}{10} = 0.9$$

Plug this value for y back into the first equation to solve for x:
$$x = 15 - 6(0.9) = 15 - 5.4 = 9.6$$

Check both equations if you have time:
$$9.6 + 6(0.9) = 9.6 + 5.4 = 15$$
$$3(9.6) - 12(0.9) = 28.8 - 10.8 = 18$$
Therefore, the solution is (9.6, 0.9).

To solve a system of equations using *elimination*, begin by rewriting both equations in standard form $Ax + By = C$. Check to see if the coefficients of one pair of like variables add to zero. If not, multiply one or both of the equations by a non-zero number to make one set of like variables add to zero. Add the two equations to solve for one of the variables. Substitute this value into one of the original equations to solve for the other variable. Check your work by substituting into the other equation. Next we will solve the same problem as above, but using the addition method.

Solve the system using elimination:
$$x + 6y = 15$$
$$3x - 12y = 18$$

If we multiply the first equation by 2, we can eliminate the y terms:
$$2x + 12y = 30$$
$$3x - 12y = 18$$

Add the equations together and solve for x:
$$5x = 48$$
$$x = \frac{48}{5} = 9.6$$

Plug the value for x back into either of the original equations and solve for y:
$$9.6 + 6y = 15$$
$$y = \frac{15 - 9.6}{6} = 0.9$$

Check both equations if you have time:
$$9.6 + 6(0.9) = 9.6 + 5.4 = 15$$
$$3(9.6) - 12(0.9) = 28.8 - 10.8 = 18$$
Therefore, the solution is (9.6, 0.9).

Polynomial Algebra

To multiply two binomials, follow the *FOIL* method. FOIL stands for:
- First: Multiply the first term of each binomial
- Outer: Multiply the outer terms of each binomial
- Inner: Multiply the inner terms of each binomial
- Last: Multiply the last term of each binomial

Using FOIL $(Ax + By)(Cx + Dy) = ACx^2 + ADxy + BCxy + BDy^2$.

> ➤ **Review Video: <u>Multiplying Terms Using the FOIL Method</u>**
> *Visit **mometrix.com/academy** and enter **Code**: 854792*

To divide polynomials, begin by arranging the terms of each polynomial in order of one variable. You may arrange in ascending or descending order, but be consistent with both polynomials. To get the first term of the quotient, divide the first term of the dividend by the first term of the divisor. Multiply the first term of the quotient by the entire divisor and subtract that product from the dividend. Repeat for the second and successive terms until you either get a remainder of zero or a remainder whose degree is less than the degree of the divisor. If the quotient has a remainder, write the answer as a mixed expression in the form: quotient $+ \frac{\text{remainder}}{\text{divisor}}$.

Rational Expressions are fractions with polynomials in both the numerator and the denominator; the value of the polynomial in the denominator cannot be equal to zero. To add or subtract rational expressions, first find the common denominator, then rewrite each fraction as an equivalent fraction with the common denominator. Finally, add or subtract the numerators to get the numerator of the answer, and keep the common denominator as the denominator of the answer. When multiplying rational expressions factor each polynomial and cancel like factors (a factor which appears in both the numerator and the denominator). Then, multiply all remaining factors in the numerator to get the numerator of the product, and multiply the remaining factors in the denominator to get the denominator of the product. Remember – cancel entire factors, not individual terms. To divide rational expressions, take the reciprocal of the divisor (the rational expression you are dividing by) and multiply by the dividend.

Below are patterns of some special products to remember: *perfect trinomial squares*, the *difference between two squares*, the *sum and difference of two cubes*, and *perfect cubes*.

- Perfect Trinomial Squares: $x^2 + 2xy + y^2 = (x + y)^2$ or $x^2 - 2xy + y^2 = (x - y)^2$
- Difference Between Two Squares: $x^2 - y^2 = (x + y)(x - y)$
- Sum of Two Cubes: $x^3 + y^3 = (x + y)(x^2 - xy + y^2)$
- Note: the second factor is NOT the same as a perfect trinomial square, so do not try to factor it further.
- Difference between Two Cubes: $x^3 - y^3 = (x - y)(x^2 + xy + y^2)$
- Again, the second factor is NOT the same as a perfect trinomial square.
- Perfect Cubes: $x^3 + 3x^2y + 3xy^2 + y^3 = (x + y)^3$ and $x^3 - 3x^2y + 3xy^2 - y^3 = (x - y)^3$

In order to *factor* a polynomial, first check for a common monomial factor. When the greatest common monomial factor has been factored out, look for patterns of special products: differences of two squares, the sum or difference of two cubes for binomial factors, or perfect trinomial squares for trinomial factors. If the factor is a trinomial but not a perfect trinomial square, look for a factorable form, such as $x^2 + (a + b)x + ab = (x + a)(x + b)$ or $(ac)x^2 + (ad + bc)x + bd = (ax + b)(cx + d)$. For factors with four terms, look for groups to factor. Once you have found the factors, write the original polynomial as the product of all the factors. Make sure all of the polynomial factors are prime. Monomial factors may be prime or composite. Check your work by multiplying the factors to make sure you get the original polynomial.

Solving Quadratic Equations

The *Quadratic Formula* is used to solve quadratic equations when other methods are more difficult. To use the quadratic formula to solve a quadratic equation, begin by rewriting the equation in standard form $ax^2 + bx + c = 0$, where a, b, and c are coefficients. Once you have identified the values of the coefficients, substitute those values into the quadratic formula $x = \frac{-b \pm \sqrt{b^2 - 4ac}}{2a}$. Evaluate the equation and simplify the expression. Again, check each root by substituting into the original equation. In the quadratic formula, the portion of the formula under the radical $(b^2 - 4ac)$ is called the *Discriminant*. If the discriminant is zero, there is only one root: zero. If the discriminant is positive, there are two different real roots. If the discriminant is negative, there are no real roots.

To solve a quadratic equation by *Factoring*, begin by rewriting the equation in standard form, if necessary. Factor the side with the variable then set each of the factors equal to zero and solve the resulting linear equations. Check your answers by substituting the roots you found into the original equation. If, when writing the equation in standard form, you have an equation in the form $x^2 + c = 0$ or $x^2 - c = 0$, set $x^2 = -c$ or $x^2 = c$ and take the square root of c. If $c = 0$, the only real root is zero. If c is positive, there are two real roots—the positive and negative square root values. If c is negative, there are no real roots because you cannot take the square root of a negative number.

> ➤ **Review Video: <u>Factoring Quadratic Equations</u>**
> *Visit **mometrix.com/academy** and enter **Code**: 336566*

To solve a quadratic equation by *Completing the Square*, rewrite the equation so that all terms containing the variable are on the left side of the equal sign, and all the constants are on the right side of the equal sign. Make sure the coefficient of the squared term is 1. If there is a coefficient with the squared term, divide each term on both sides of the equal side by that number. Next, work with the coefficient of the single-variable term. Square half of this coefficient, and add that value to both sides. Now you can factor the left side (the side containing the variable) as the square of a binomial. $x^2 + 2ax + a^2 = C \Rightarrow (x + a)^2 = C$, where x is the variable, and a and C are constants. Take the square root of both sides and solve for the variable. Substitute the value of the variable in the original problem to check your work.

Matrix Basics
A **matrix** (plural: matrices) is a rectangular array of numbers or variables, often called **elements**, which are arranged in columns and rows. A matrix is generally represented by a capital letter, with its elements represented by the corresponding lowercase letter with two subscripts indicating the row and column of the element. For example, n_{ab} represents the element in row a column b of matrix N.

$$N = \begin{bmatrix} n_{11} & n_{12} & n_{13} \\ n_{21} & n_{22} & n_{23} \end{bmatrix}$$

A matrix can be described in terms of the number of rows and columns it contains in the format $a \times b$, where a is the number of rows and b is the number of columns. The matrix shown above is a

2×3 matrix. Any $a \times b$ matrix where $a = b$ is a square matrix. A **vector** is a matrix that has exactly one column (**column vector**) or exactly one row (**row vector**).

The **main diagonal** of a matrix is the set of elements on the diagonal from the top left to the bottom right of a matrix. Because of the way it is defined, only square matrices will have a main diagonal. For the matrix shown below, the main diagonal consists of the elements $n_{11}, n_{22}, n_{33}, n_{44}$.

$$\begin{bmatrix} n_{11} & n_{12} & n_{13} & n_{14} \\ n_{21} & n_{22} & n_{23} & n_{24} \\ n_{31} & n_{32} & n_{33} & n_{34} \\ n_{41} & n_{42} & n_{43} & n_{44} \end{bmatrix}$$

A 3×4 matrix such as the one shown below would not have a main diagonal because there is no straight line of elements between the top left corner and the bottom right corner that joins the elements.

$$\begin{bmatrix} n_{11} & n_{12} & n_{13} & n_{14} \\ n_{21} & n_{22} & n_{23} & n_{24} \\ n_{31} & n_{32} & n_{33} & n_{34} \end{bmatrix}$$

A **diagonal matrix** is a square matrix that has a zero for every element in the matrix except the elements on the main diagonal. All the elements on the main diagonal must be nonzero numbers.

$$\begin{bmatrix} n_{11} & 0 & 0 & 0 \\ 0 & n_{22} & 0 & 0 \\ 0 & 0 & n_{33} & 0 \\ 0 & 0 & 0 & n_{44} \end{bmatrix}$$

If every element on the main diagonal of a diagonal matrix is equal to one, the matrix is called an **identity matrix**. The identity matrix is often represented by the letter I.

$$I = \begin{bmatrix} 1 & 0 & 0 & 0 \\ 0 & 1 & 0 & 0 \\ 0 & 0 & 1 & 0 \\ 0 & 0 & 0 & 1 \end{bmatrix}$$

A **zero matrix** is a matrix that has zero as the value for every element in the matrix.

$$\begin{bmatrix} 0 & 0 & 0 & 0 \\ 0 & 0 & 0 & 0 \\ 0 & 0 & 0 & 0 \\ 0 & 0 & 0 & 0 \end{bmatrix}$$

The zero matrix is the *identity for matrix addition*. Do not confuse the zero matrix with the identity matrix.

The **negative of a matrix** is also known as the additive inverse of a matrix. If matrix N is the given matrix, then matrix $-N$ is its negative. This means that every element n_{ab} is equal to $-n_{ab}$ in the negative. To find the negative of a given matrix, change the sign of every element in the matrix and keep all elements in their original corresponding positions in the matrix.

If two matrices have the same order and all corresponding elements in the two matrices are the same, then the two matrices are **equal matrices**.

A matrix N may be **transposed** to matrix N^T by changing all rows into columns and changing all columns into rows. The easiest way to accomplish this is to swap the positions of the row and column notations for each element. For example, suppose the element in the second row of the third column of matrix N is $n_{23} = 6$. In the transposed matrix N^T, the transposed element would be $n_{32} = 6$, and it would be placed in the third row of the second column.

$$N = \begin{bmatrix} 1 & 2 & 3 \\ 4 & 5 & 6 \end{bmatrix}; \ N^T = \begin{bmatrix} 1 & 4 \\ 2 & 5 \\ 3 & 6 \end{bmatrix}$$

To quickly transpose a matrix by hand, begin with the first column and rewrite a new matrix with those same elements in the same order in the first row. Write the elements from the second column of the original matrix in the second row of the transposed matrix. Continue this process until all columns have been completed. If the original matrix is identical to the transposed matrix, the matrices are symmetric.

The **determinant** of a matrix is a scalar value that is calculated by taking into account all the elements of a square matrix. A determinant only exists for square matrices. Finding the determinant of a 2×2 matrix is as simple as remembering a simple equation. For a 2×2 matrix $M = \begin{bmatrix} m_{11} & m_{12} \\ m_{21} & m_{22} \end{bmatrix}$, the determinant is obtained by the equation $|M| = m_{11}m_{22} - m_{12}m_{21}$. Anything larger than 2×2 requires multiple steps. Take matrix $N = \begin{bmatrix} a & b & c \\ d & e & f \\ g & h & j \end{bmatrix}$. The determinant of N is calculated as $|N| = a \begin{vmatrix} e & f \\ h & j \end{vmatrix} - b \begin{vmatrix} d & f \\ g & j \end{vmatrix} + c \begin{vmatrix} d & e \\ g & h \end{vmatrix}$ or $|N| = a(ej - fh) - b(dj - fg) + c(dh - eg)$.

There is a shortcut for 3×3 matrices: add the products of each unique set of elements diagonally left-to-right and subtract the products of each unique set of elements diagonally right-to-left. In matrix N, the left-to-right diagonal elements are (a, e, j), (b, f, g), and (c, d, h). The right-to-left diagonal elements are (a, f, h), (b, d, j), and (c, e, g). $\det(N) = aej + bfg + cdh - afh - bdj - ceg$.

Calculating the determinants of matrices larger than 3×3 is rarely, if ever, done by hand.

The **inverse** of a matrix M is the matrix that, when multiplied by matrix M, yields a product that is the identity matrix. Multiplication of matrices will be explained in greater detail shortly. Not all matrices have inverses. Only a square matrix whose determinant is not zero has an inverse. If a matrix has an inverse, that inverse is unique to that matrix. For any matrix M that has an inverse, the inverse is represented by the symbol M^{-1}. To calculate the inverse of a 2×2 square matrix, use the following pattern:

$$M = \begin{bmatrix} m_{11} & m_{12} \\ m_{21} & m_{22} \end{bmatrix}; \ M^{-1} = \begin{bmatrix} \dfrac{m_{22}}{|M|} & \dfrac{-m_{12}}{|M|} \\ \dfrac{-m_{21}}{|M|} & \dfrac{m_{11}}{|M|} \end{bmatrix}$$

Another way to find the inverse of a matrix by hand is use an augmented matrix and elementary row operations. An **augmented matrix** is formed by appending the entries from one matrix onto the end of another. For example, given a 2×2 invertible matrix $N = \begin{bmatrix} a & b \\ c & d \end{bmatrix}$, you can find the inverse N^{-1} by creating an augmented matrix by appending a 2×2 identity matrix: $\begin{bmatrix} a & b & | & 1 & 0 \\ c & d & | & 0 & 1 \end{bmatrix}$. To find the inverse of the original 2×2 matrix, perform elementary row operations to convert the original matrix on the left to an identity matrix: $\begin{bmatrix} 1 & 0 & | & e & f \\ 0 & 1 & | & g & h \end{bmatrix}$.

Elementary row operations include multiplying a row by a non-zero scalar, adding scalar multiples of two rows, or some combination of these. For instance, the first step might be to multiply the second row by $\frac{b}{d}$ and then subtract it from the first row to make its second column a zero. The end result is that the 2×2 section on the right will become the inverse of the original matrix: $N^{-1} = \begin{bmatrix} e & f \\ g & h \end{bmatrix}$.

Calculating the inverse of any matrix larger than 2×2 is cumbersome and using a graphing calculator is recommended.

Basic Operations with Matrices

There are two categories of basic operations with regard to matrices: operations between a matrix and a scalar, and operations between two matrices.

Scalar Operations
A scalar being added to a matrix is treated as though it were being added to each element of the matrix:

$$M + 4 = \begin{bmatrix} m_{11} + 4 & m_{12} + 4 \\ m_{21} + 4 & m_{22} + 4 \end{bmatrix}$$

The same is true for the other three operations.

Subtraction:

$$M - 4 = \begin{bmatrix} m_{11} - 4 & m_{12} - 4 \\ m_{21} - 4 & m_{22} - 4 \end{bmatrix}$$

Multiplication:

$$M \times 4 = \begin{bmatrix} m_{11} \times 4 & m_{12} \times 4 \\ m_{21} \times 4 & m_{22} \times 4 \end{bmatrix}$$

Division:

$$M \div 4 = \begin{bmatrix} m_{11} \div 4 & m_{12} \div 4 \\ m_{21} \div 4 & m_{22} \div 4 \end{bmatrix}$$

Matrix Addition and Subtraction
All four of the basic operations can be used with operations between matrices (although division is usually discarded in favor of multiplication by the inverse), but there are restrictions on the situations in which they can be used. Matrices that meet all the qualifications for a given operation are called **conformable matrices**. However, conformability is specific to the operation; two matrices that are conformable for addition are not necessarily conformable for multiplication.

For two matrices to be conformable for addition or subtraction, they must be of the same dimension; otherwise the operation is not defined. If matrix M is a 3×2 matrix and matrix N is a 2×3 matrix, the operations $M + N$ and $M - N$ are meaningless. If matrices M and N are the same size, the operation is as simple as adding or subtracting all of the corresponding elements:

$$\begin{bmatrix} m_{11} & m_{12} \\ m_{21} & m_{22} \end{bmatrix} + \begin{bmatrix} n_{11} & n_{12} \\ n_{21} & n_{22} \end{bmatrix} = \begin{bmatrix} m_{11} + n_{11} & m_{12} + n_{12} \\ m_{21} + n_{21} & m_{22} + n_{22} \end{bmatrix}$$

$$\begin{bmatrix} m_{11} & m_{12} \\ m_{21} & m_{22} \end{bmatrix} - \begin{bmatrix} n_{11} & n_{12} \\ n_{21} & n_{22} \end{bmatrix} = \begin{bmatrix} m_{11} - n_{11} & m_{12} - n_{12} \\ m_{21} - n_{21} & m_{22} - n_{22} \end{bmatrix}$$

The result of addition or subtraction is a matrix of the same dimension as the two original matrices involved in the operation.

Multiplication

The first thing it is necessary to understand about matrix multiplication is that it is not commutative. In scalar multiplication, the operation is commutative, meaning that $a \times b = b \times a$. For matrix multiplication, this is not the case: $A \times B \neq B \times A$. The terminology must be specific when describing matrix multiplication. The operation $A \times B$ can be described as A multiplied (or **post-multiplied**) by B, or B **pre-multiplied** by A.

For two matrices to be conformable for multiplication, they need not be of the same dimension, but specific dimensions must correspond. Taking the example of two matrices M and N to be multiplied $M \times N$, matrix M must have the same number of columns as matrix N has rows. Put another way, if matrix M has the dimensions $a \times b$ and matrix N has the dimensions $c \times d$, b must equal c if the two matrices are to be conformable for this multiplication. The matrix that results from the multiplication will have the dimensions $a \times d$. If a and d are both equal to 1, the product is simply a scalar. Square matrices of the same dimensions are always conformable for multiplication, and their product is always a matrix of the same size.

The simplest type of matrix multiplication is a 1×2 matrix (a row vector) times a 2×1 matrix (a column vector). These will multiply in the following way:

$$[m_{11} \quad m_{12}] \times \begin{bmatrix} n_{11} \\ n_{21} \end{bmatrix} = m_{11}n_{11} + m_{12}n_{21}$$

The two matrices are conformable for multiplication because matrix M has the same number of columns as matrix N has rows. Because the other dimensions are both 1, the result is a scalar. Expanding our matrices to 1×3 and 3×1, the process is the same:

$$[m_{11} \quad m_{12} \quad m_{13}] \times \begin{bmatrix} n_{11} \\ n_{21} \\ n_{31} \end{bmatrix} = m_{11}n_{11} + m_{12}n_{21} + m_{13}n_{31}$$

Once again, the result is a scalar. This type of basic matrix multiplication is the building block for the multiplication of larger matrices.

To multiply larger matrices, treat each **row from the first matrix** and each **column from the second matrix** as individual vectors and follow the pattern for multiplying vectors. The scalar value found from multiplying the first row vector by the first column vector is placed in the first row, first column of the new matrix. The scalar value found from multiplying the second row vector by the first column vector is placed in the second row, first column of the new matrix. Continue this pattern until each row of the first matrix has been multiplied by each column of the second vector.

Below is an example of the multiplication of a 3×2 matrix and a 2×3 matrix.

$$\begin{bmatrix} m_{11} & m_{12} \\ m_{21} & m_{22} \\ m_{31} & m_{32} \end{bmatrix} \times \begin{bmatrix} n_{11} & n_{12} & n_{13} \\ n_{21} & n_{22} & n_{23} \end{bmatrix} = \begin{bmatrix} m_{11}n_{11} + m_{12}n_{21} & m_{11}n_{12} + m_{12}n_{22} & m_{11}n_{13} + m_{12}n_{23} \\ m_{21}n_{11} + m_{22}n_{21} & m_{21}n_{12} + m_{22}n_{22} & m_{21}n_{13} + m_{22}n_{23} \\ m_{31}n_{11} + m_{32}n_{21} & m_{31}n_{12} + m_{32}n_{22} & m_{31}n_{13} + m_{32}n_{23} \end{bmatrix}$$

The result is a 3×3 matrix. If the operation were done in reverse ($N \times M$), the result would be a 2×2 matrix.

$$\begin{bmatrix} n_{11} & n_{12} & n_{13} \\ n_{21} & n_{22} & n_{23} \end{bmatrix} \times \begin{bmatrix} m_{11} & m_{12} \\ m_{21} & m_{22} \\ m_{31} & m_{32} \end{bmatrix} = \begin{bmatrix} m_{11}n_{11} + m_{21}n_{12} + m_{31}n_{13} & m_{12}n_{11} + m_{22}n_{12} + m_{32}n_{13} \\ m_{11}n_{21} + m_{21}n_{22} + m_{31}n_{23} & m_{12}n_{21} + m_{22}n_{22} + m_{32}n_{23} \end{bmatrix}$$

Solving Systems of Equations

Matrices can be used to represent the coefficients of a system of linear equations and can be very useful in solving those systems. Take for instance three equations with three variables:

$$a_1 x + b_1 y + c_1 z = d_1$$
$$a_2 x + b_2 y + c_2 z = d_2$$
$$a_3 x + b_3 y + c_3 z = d_3$$

where all a, b, c, and d are known constants.

To solve this system, define three matrices:

$$A = \begin{bmatrix} a_1 & b_1 & c_1 \\ a_2 & b_2 & c_2 \\ a_3 & b_3 & c_3 \end{bmatrix}; \quad D = \begin{bmatrix} d_1 \\ d_2 \\ d_3 \end{bmatrix}; \quad X = \begin{bmatrix} x \\ y \\ z \end{bmatrix}$$

The three equations in our system can be fully represented by a single matrix equation:

$$AX = D$$

We know that the identity matrix times X is equal to X, and we know that any matrix multiplied by its inverse is equal to the identity matrix.

$$A^{-1}AX = IX = X; \text{thus } X = A^{-1}D$$

Our goal then is to find the inverse of A, or A^{-1}. Once we have that, we can pre-multiply matrix D by A^{-1} (post-multiplying here is an undefined operation) to find matrix X.

Systems of equations can also be solved using the transformation of an augmented matrix in a process similar to that for finding a matrix inverse. Begin by arranging each equation of the system in the following format:

$$a_1 x + b_1 y + c_1 z = d_1$$
$$a_2 x + b_2 y + c_2 z = d_2$$
$$a_3 x + b_3 y + c_3 z = d_3$$

Define matrices A and D and combine them into augmented matrix A_a:

$$A = \begin{bmatrix} a_1 & b_1 & c_1 \\ a_2 & b_2 & c_2 \\ a_3 & b_3 & c_3 \end{bmatrix}; D = \begin{bmatrix} d_1 \\ d_2 \\ d_3 \end{bmatrix}; A_a = \begin{bmatrix} a_1 & b_1 & c_1 & d_1 \\ a_2 & b_2 & c_2 & d_2 \\ a_3 & b_3 & c_3 & d_3 \end{bmatrix}$$

To solve the augmented matrix and the system of equations, use elementary row operations to form an identity matrix in the first 3×3 section. When this is complete, the values in the last column are the solutions to the system of equations:

$$\begin{bmatrix} 1 & 0 & 0 & x \\ 0 & 1 & 0 & y \\ 0 & 0 & 1 & z \end{bmatrix}$$

If an identity matrix is not possible, the system of equations has no unique solution. Sometimes only a partial solution will be possible. The following are partial solutions you may find:

$$\begin{bmatrix} 1 & 0 & k_1 & x_0 \\ 0 & 1 & k_2 & y_0 \\ 0 & 0 & 0 & 0 \end{bmatrix}$$ gives the non-unique solution $x = x_0 - k_1 z$; $y = y_0 - k_2 z$

$$\begin{bmatrix} 1 & j_1 & k_1 & x_0 \\ 0 & 0 & 0 & 0 \\ 0 & 0 & 0 & 0 \end{bmatrix}$$ gives the non-unique solution $x = x_0 - j_1 y - k_1 z$

This process can be used to solve systems of equations with any number of variables, but three is the upper limit for practical purposes. Anything more ought to be done with a graphing calculator.

Geometric Transformations

The four geometric transformations are translations, reflections, rotations, and dilations. When geometric transformations are expressed as matrices, the process of performing the transformations is simplified. For calculations of the geometric transformations of a planar figure, make a $2 \times n$ matrix, where n is the number of vertices in the planar figure. Each column represents the rectangular coordinates of one vertex of the figure, with the top row containing the values of the x-coordinates and the bottom row containing the values of the y-coordinates. For example, given a planar triangular figure with coordinates (x_1, y_1), (x_2, y_2), and (x_3, y_3), the corresponding matrix is $\begin{bmatrix} x_1 & x_2 & x_3 \\ y_1 & y_2 & y_3 \end{bmatrix}$. You can then perform the necessary transformations on this matrix to determine the coordinates of the resulting figure.

Translation
A translation moves a figure along the x-axis, the y-axis, or both axes without changing the size or shape of the figure. To calculate the new coordinates of a planar figure following a translation, set up a matrix of the coordinates and a matrix of the translation values and add the two matrices.

$$\begin{bmatrix} h & h & h \\ v & v & v \end{bmatrix} + \begin{bmatrix} x_1 & x_2 & x_3 \\ y_1 & y_2 & y_3 \end{bmatrix} = \begin{bmatrix} h + x_1 & h + x_2 & h + x_3 \\ v + y_1 & v + y_2 & v + y_3 \end{bmatrix}$$

where h is the number of units the figure is moved along the x-axis (horizontally) and v is the number of units the figure is moved along the y-axis (vertically).

Reflection

To find the reflection of a planar figure over the *x*-axis, set up a matrix of the coordinates of the vertices and pre-multiply the matrix by the 2×2 matrix $\begin{bmatrix} 1 & 0 \\ 0 & -1 \end{bmatrix}$ so that $\begin{bmatrix} 1 & 0 \\ 0 & -1 \end{bmatrix}\begin{bmatrix} x_1 & x_2 & x_3 \\ y_1 & y_2 & y_3 \end{bmatrix} = \begin{bmatrix} x_1 & x_2 & x_3 \\ -y_1 & -y_2 & -y_3 \end{bmatrix}$. To find the reflection of a planar figure over the *y*-axis, set up a matrix of the coordinates of the vertices and pre-multiply the matrix by the 2×2 matrix $\begin{bmatrix} -1 & 0 \\ 0 & 1 \end{bmatrix}$ so that $\begin{bmatrix} -1 & 0 \\ 0 & 1 \end{bmatrix}\begin{bmatrix} x_1 & x_2 & x_3 \\ y_1 & y_2 & y_3 \end{bmatrix} = \begin{bmatrix} -x_1 & -x_2 & -x_3 \\ y_1 & y_2 & y_3 \end{bmatrix}$. To find the reflection of a planar figure over the line $y = x$, set up a matrix of the coordinates of the vertices and pre-multiply the matrix by the 2×2 matrix $\begin{bmatrix} 0 & 1 \\ 1 & 0 \end{bmatrix}$ so that $\begin{bmatrix} 0 & 1 \\ 1 & 0 \end{bmatrix}\begin{bmatrix} x_1 & x_2 & x_3 \\ y_1 & y_2 & y_3 \end{bmatrix} = \begin{bmatrix} y_1 & y_2 & y_3 \\ x_1 & x_2 & x_3 \end{bmatrix}$. Remember that the order of multiplication is important when multiplying matrices. The commutative property does not apply.

Rotation

To find the coordinates of the figure formed by rotating a planar figure about the origin θ degrees in a counterclockwise direction, set up a matrix of the coordinates of the vertices and pre-multiply the matrix by the 2×2 matrix $\begin{bmatrix} \cos\theta & \sin\theta \\ -\sin\theta & \cos\theta \end{bmatrix}$. For example, if you want to rotate a figure 90° clockwise around the origin, you would have to convert the degree measure to 270° counterclockwise and solve the 2×2 matrix you have set as the pre-multiplier: $\begin{bmatrix} \cos 270° & \sin 270° \\ -\sin 270° & \cos 270° \end{bmatrix} = \begin{bmatrix} 0 & -1 \\ 1 & 0 \end{bmatrix}$. Use this as the pre-multiplier for the matrix $\begin{bmatrix} x_1 & x_2 & x_3 \\ y_1 & y_2 & y_3 \end{bmatrix}$ and solve to find the new coordinates.

Dilation

To find the dilation of a planar figure by a scale factor of k, set up a matrix of the coordinates of the vertices of the planar figure and pre-multiply the matrix by the 2×2 matrix $\begin{bmatrix} k & 0 \\ 0 & k \end{bmatrix}$ so that $\begin{bmatrix} k & 0 \\ 0 & k \end{bmatrix}\begin{bmatrix} x_1 & x_2 & x_3 \\ y_1 & y_2 & y_3 \end{bmatrix} = \begin{bmatrix} kx_1 & kx_2 & kx_3 \\ ky_1 & ky_2 & ky_3 \end{bmatrix}$. This is effectively the same as multiplying the matrix by the scalar k, but the matrix equation would still be necessary if the figure were being dilated by different factors in vertical and horizontal directions. The scale factor k will be greater than 1 if the figure is being enlarged, and between 0 and 1 if the figure is being shrunk. Again, remember that when multiplying matrices, the order of the matrices is important. The commutative property does not apply, and the matrix with the coordinates of the figure must be the second matrix.

Functions

Functions

A function is an equation that has exactly one value of output variable (dependent variable) for each value of the input variable (independent variable). The set of all values for the input variable (here assumed to be x) is the domain of the function, and the set of all corresponding values of output variable (here assumed to be y) is the range of the function. When looking at a graph of an equation, the easiest way to determine if the equation is a function or not is to conduct the vertical line test. If a vertical line drawn through any value of x crosses the graph in more than one place, the equation is not a function.

In functions with the notation $f(x)$, the value substituted for x in the equation is called the argument. The domain is the set of all values for x in a function. Unless otherwise given, assume the domain is the set of real numbers that will yield real numbers for the range. This is the domain of definition.

The graph of a function is the set of all ordered pairs (x, y) that satisfy the equation of the function. The points that have zero as the value for y are called the zeros of the function. These are also the x-intercepts, because that is the point at which the graph crosses, or intercepts, the x-axis. The points that have zero as the value for x are the y-intercepts because that is where the graph crosses the y-axis.

> ➤ **Review Video: Basics of Functions**
> *Visit **mometrix.com/academy** and enter **Code**: **822500***

Any time there are vertical asymptotes or holes in a graph, such that the complete graph cannot be drawn as one continuous line, a graph is said to have discontinuities. Examples would include the graphs of hyperbolas that are functions, and the function $f(x) = \tan x$.

> ➤ **Review Video: Graphs of Functions**
> *Visit **mometrix.com/academy** and enter **Code**: **492785***

Manipulation of Functions

Horizontal and vertical shift occur when values are added to or subtracted from the x or y values, respectively.

If a constant is added to the y portion of each point, the graph shifts up. If a constant is subtracted from the y portion of each point, the graph shifts down. This is represented by the expression $f(x) \pm k$, where k is a constant.

If a constant is added to the x portion of each point, the graph shifts left. If a constant is subtracted from the x portion of each point, the graph shifts right. This is represented by the expression $f(x \pm k)$, where k is a constant.

Stretch, compression, and reflection occur when different parts of a function are multiplied by different groups of constants. If the function as a whole is multiplied by a real number constant greater than 1, $(k \times f(x))$, the graph is stretched vertically. If k in the previous equation is greater than zero but less than 1, the graph is compressed vertically. If k is less than zero, the graph is reflected about the x-axis, in addition to being either stretched or compressed vertically if k is less than or greater than -1, respectively. If instead, just the x-term is multiplied by a constant greater than 1 $(f(k \times x))$, the graph is compressed horizontally. If k in the previous equation is greater than zero but less than 1, the graph is stretched horizontally. If k is less than zero, the graph is reflected about the y-axis, in addition to being either stretched or compressed horizontally if k is greater than or less than -1, respectively.

Classification of Functions

There are many different ways to classify functions based on their structure or behavior. Listed here are a few common classifications.

Constant functions are given by the equation y = b or $f(x) = b$, where b is a real number. There is no independent variable present in the equation, so the function has a constant value for all x. The graph of a constant function is a horizontal line of slope 0 that is positioned b units from the x-axis. If b is positive, the line is above the x-axis; if b is negative, the line is below the x-axis.

Identity functions are identified by the equation y = x or $f(x) = x$, where every value of y is equal to its corresponding value of x. The only zero is the point (0, 0). The graph is a diagonal line with slope 1.

In **linear functions**, the value of the function changes in direct proportion to x. The rate of change, represented by the slope on its graph, is constant throughout. The standard form of a linear equation is $ax + by = c$, where a, b, and c are real numbers. As a function, this equation is commonly written as $y = mx + b$ or $f(x) = mx + b$. This is known as the slope-intercept form, because the coefficients give the slope of the graphed function (m) and its y-intercept (b). Solve the equation $mx + b = 0$ for x to get $x = -\frac{b}{m}$, which is the only zero of the function. The domain and range are both the set of all real numbers.

A **polynomial function** is a function with multiple terms and multiple powers of x, such as
$$f(x) = a_n x^n + a_{n-1} x^{n-1} + a_{n-2} x^{n-2} + \cdots + a_1 x + a_0$$

where n is a non-negative integer that is the highest exponent in the polynomial, and $a_n \neq 0$. The domain of a polynomial function is the set of all real numbers. If the greatest exponent in the polynomial is even, the polynomial is said to be of even degree and the range is the set of real numbers that satisfy the function. If the greatest exponent in the polynomial is odd, the polynomial is said to be odd and the range, like the domain, is the set of all real numbers.

> ➢ **Review Video:** <u>Simplifying Rational Polynomial Functions</u>
> *Visit **mometrix.com/academy** and enter **Code**: **893868***

A **quadratic function** is a polynomial function that follows the equation pattern $y = ax^2 + bx + c$, or $f(x) = ax^2 + bx + c$, where a, b, and c are real numbers and $a \neq 0$. The domain of a quadratic function is the set of all real numbers. The range is also real numbers, but only those in the subset of the domain that satisfy the equation. The root(s) of any quadratic function can be found by plugging the values of a, b, and c into the **quadratic formula**:

$$x = \frac{-b \pm \sqrt{b^2 - 4ac}}{2a}$$

If the expression $b^2 - 4ac$ is negative, you will instead find complex roots.

A quadratic function has a parabola for its graph. In the equation $f(x) = ax^2 + bx + c$, if a is positive, the parabola will open upward. If a is negative, the parabola will open downward. The axis of symmetry is a vertical line that passes through the vertex. To determine whether or not the parabola will intersect the x-axis, check the number of real roots. An equation with two real roots will cross the x-axis twice. An equation with one real root will have its vertex on the x-axis. An equation with no real roots will not contact the x-axis.

➤ **Review Video:** <u>Changing Constants in Graphs of Functions: Quadratic Equations</u>
*Visit **mometrix.com/academy** and enter **Code**: 476276*

A **rational function** is a function that can be constructed as a ratio of two polynomial expressions: $f(x) = \frac{p(x)}{q(x)}$, where $p(x)$ and $q(x)$ are both polynomial expressions and $q(x) \neq 0$. The domain is the set of all real numbers, except any values for which $q(x) = 0$. The range is the set of real numbers that satisfies the function when the domain is applied. When you graph a rational function, you will have vertical asymptotes wherever $q(x) = 0$. If the polynomial in the numerator is of lesser degree than the polynomial in the denominator, the x-axis will also be a horizontal asymptote. If the numerator and denominator have equal degrees, there will be a horizontal asymptote not on the x-axis. If the degree of the numerator is exactly one greater than the degree of the denominator, the graph will have an oblique, or diagonal, asymptote. The asymptote will be along the line $y = \frac{p_n}{q_{n-1}} x + \frac{p_{n-1}}{q_{n-1}}$, where p_n and q_{n-1} are the coefficients of the highest degree terms in their respective polynomials.

A **square root function** is a function that contains a radical and is in the format $f(x) = \sqrt{ax + b}$. The domain is the set of all real numbers that yields a positive radicand or a radicand equal to zero. Because square root values are assumed to be positive unless otherwise identified, the range is all real numbers from zero to infinity. To find the zero of a square root function, set the radicand equal to zero and solve for x. The graph of a square root function is always to the right of the zero and always above the x-axis.

An **absolute value function** is in the format $f(x) = |ax + b|$. Like other functions, the domain is the set of all real numbers. However, because absolute value indicates positive numbers, the range is limited to positive real numbers. To find the zero of an absolute value function, set the portion inside the absolute value sign equal to zero and solve for x.

An absolute value function is also known as a piecewise function because it must be solved in pieces – one for if the value inside the absolute value sign is positive, and one for if the value is negative. The function can be expressed as

$$f(x) = \begin{cases} ax + b & \text{if } ax + b \geq 0 \\ -(ax + b) & \text{if } ax + b < 0 \end{cases}$$

This will allow for an accurate statement of the range.

Exponential functions are equations that have the format $y = b^x$, where base $b > 0$ and $b \neq 1$. The exponential function can also be written $f(x) = b^x$.

Logarithmic functions are equations that have the format $y = \log_b x$ or $f(x) = \log_b x$. The base b may be any number except one; however, the most common bases for logarithms are base 10 and base e. The log base e is known the natural logarithm, or ln, expressed by the function $f(x) = \ln x$.

Any logarithm that does not have an assigned value of b is assumed to be base 10: $\log x = \log_{10} x$. Exponential functions and logarithmic functions are related in that one is the inverse of the other. If $f(x) = b^x$, then $f^{-1}(x) = \log_b x$. This can perhaps be expressed more clearly by the two equations: $y = b^x$ and $x = \log_b y$.

The following properties apply to logarithmic expressions:

$$\log_b 1 = 0$$
$$\log_b b = 1$$
$$\log_b b^p = p$$
$$\log_b MN = \log_b M + \log_b N$$
$$\log_b \frac{M}{N} = \log_b M - \log_b N$$
$$\log_b M^p = p \log_b M$$

In a **one-to-one function**, each value of x has exactly one value for y (this is the definition of a function) *and* each value of y has exactly one value for x. While the vertical line test will determine if a graph is that of a function, the horizontal line test will determine if a function is a one-to-one function. If a horizontal line drawn at any value of y intersects the graph in more than one place, the graph is not that of a one-to-one function. Do not make the mistake of using the horizontal line test exclusively in determining if a graph is that of a one-to-one function. A one-to-one function must pass both the vertical line test and the horizontal line test. One-to-one functions are also **invertible functions**.

A **monotone function** is a function whose graph either constantly increases or constantly decreases. Examples include the functions $f(x) = x$, $f(x) = -x$, or $f(x) = x^3$.

An **even function** has a graph that is symmetric with respect to the y-axis and satisfies the equation $f(x) = f(-x)$. Examples include the functions $f(x) = x^2$ and $f(x) = ax^n$, where a is any real number and n is a positive even integer.

An **odd function** has a graph that is symmetric with respect to the origin and satisfies the equation $f(x) = -f(-x)$. Examples include the functions $f(x) = x^3$ and $f(x) = ax^n$, where a is any real number and n is a positive odd integer.

Algebraic functions are those that exclusively use polynomials and roots. These would include polynomial functions, rational functions, square root functions, and all combinations of these functions, such as polynomials as the radicand. These combinations may be joined by addition, subtraction, multiplication, or division, but may not include variables as exponents.

Transcendental functions are all functions that are non-algebraic. Any function that includes logarithms, trigonometric functions, variables as exponents, or any combination that includes any of these is not algebraic in nature, even if the function includes polynomials or roots.

Related Concepts

According to the **Fundamental Theorem of Algebra**, every non-constant, single variable polynomial has exactly as many roots as the polynomial's highest exponent. For example, if x^4 is the largest exponent of a term, the polynomial will have exactly 4 roots. However, some of these roots may have multiplicity or be non-real numbers. For instance, in the polynomial function $f(x) = x^4 - 4x + 3$, the only real roots are 1 and -1. The root 1 has multiplicity of 2 and there is one non-real root $(-1 - \sqrt{2}i)$.

The **Remainder Theorem** is useful for determining the remainder when a polynomial is divided by a binomial. The Remainder Theorem states that if a polynomial function $f(x)$ is divided by a binomial $x - a$, where a is a real number, the remainder of the division will be the value of $f(a)$. If $f(a) = 0$, then a is a root of the polynomial.

The **Factor Theorem** is related to the Remainder Theorem and states that if $f(a) = 0$ then $(x - a)$ is a factor of the function.

According to the **Rational Root Theorem,** any rational root of a polynomial function $f(x) = a_n x^n + a_{n-1} x^{n-1} + \cdots + a_1 x + a_0$ with integer coefficients will, when reduced to its lowest terms, be a positive or negative fraction such that the numerator is a factor of a_0 and the denominator is a factor of a_n. For instance, if the polynomial function $f(x) = x^3 + 3x^2 - 4$ has any rational roots, the numerators of those roots can only be factors of 4 (1, 2, 4), and the denominators can only be factors of 1 (1). The function in this example has roots of 1 $\left(\text{or } \frac{1}{1} \right)$ and -2 $\left(\text{or } -\frac{2}{1} \right)$.

Variables that vary directly are those that either both increase at the same rate or both decrease at the same rate. For example, in the functions $f(x) = kx$ or $f(x) = kx^n$, where k and n are positive, the value of $f(x)$ increases as the value of x increases and decreases as the value of x decreases.

Variables that vary inversely are those where one increases while the other decreases. For example, in the functions $f(x) = \frac{k}{x}$ or $f(x) = \frac{k}{x^n}$ where k is a positive constant, the value of y increases as the value of x decreases, and the value of y decreases as the value of x increases.

In both cases, k is the constant of variation.

Applying the Basic Operations to Functions

For each of the basic operations, we will use these functions as examples: $f(x) = x^2$ and $g(x) = x$.

To find the sum of two functions f and g, assuming the domains are compatible, simply add the two functions together: $(f + g)(x) = f(x) + g(x) = x^2 + x$

To find the difference of two functions f and g, assuming the domains are compatible, simply subtract the second function from the first: $(f - g)(x) = f(x) - g(x) = x^2 - x$.

To find the product of two functions f and g, assuming the domains are compatible, multiply the two functions together: $(f \cdot g)(x) = f(x) \cdot g(x) = x^2 \cdot x = x^3$.

To find the quotient of two functions f and g, assuming the domains are compatible, divide the first function by the second: $\frac{f}{g}(x) = \frac{f(x)}{g(x)} = \frac{x^2}{x} = x \, ; x \neq 0$.

The example given in each case is fairly simple, but on a given problem, if you are looking only for the value of the sum, difference, product or quotient of two functions at a particular x-value, it may be simpler to solve the functions individually and then perform the given operation using those values.

The composite of two functions f and g, written as $(f \circ g)(x)$ simply means that the output of the second function is used as the input of the first. This can also be written as $f(g(x))$. In general, this can be solved by substituting $g(x)$ for all instances of x in $f(x)$ and simplifying. Using the example functions $f(x) = x^2 - x + 2$ and $g(x) = x + 1$, we can find that $(f \circ g)(x)$ or $f(g(x))$ is equal to $f(x + 1) = (x + 1)^2 - (x + 1) + 2$, which simplifies to $x^2 + x + 2$.

It is important to note that $(f \circ g)(x)$ is not necessarily the same as $(g \circ f)(x)$. The process is not commutative like addition or multiplication expressions. If $(f \circ g)(x)$ does equal $(g \circ f)(x)$, the two functions are inverses of each other.

Basic Trigonometric Functions

The three basic trigonometric functions are sine, cosine, and tangent.

Sine

The sine (sin) function has a period of 360° or 2π radians. This means that its graph makes one complete cycle every 360° or 2π. Because $\sin 0 = 0$, the graph of $y = \sin x$ begins at the origin, with the x-axis representing the angle measure, and the y-axis representing the sine of the angle. The graph of the sine function is a smooth curve that begins at the origin, peaks at the point $\left(\frac{\pi}{2}, 1\right)$, crosses the x-axis at $(\pi, 0)$, has its lowest point at $\left(\frac{3\pi}{2}, -1\right)$, and returns to the x-axis to complete one cycle at $(2\pi, 0)$.

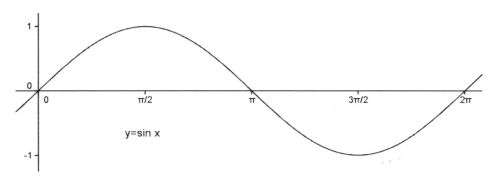

Cosine

The cosine (cos) function also has a period of 360° or 2π radians, which means that its graph also makes one complete cycle every 360° or 2π. Because $\cos 0° = 1$, the graph of $y = \cos x$ begins at the point $(0, 1)$, with the x-axis representing the angle measure, and the y-axis representing the cosine of the angle. The graph of the cosine function is a smooth curve that begins at the point $(0, 1)$, crosses the x-axis at the point $\left(\frac{\pi}{2}, 0\right)$, has its lowest point at $(\pi, -1)$, crosses the x-axis again at the point $\left(\frac{3\pi}{2}, 0\right)$, and returns to a peak at the point $(2\pi, 1)$ to complete one cycle.

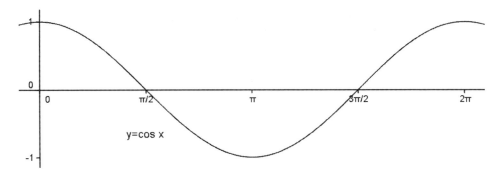

➤ **Review Video: Cosine**
*Visit **mometrix.com/academy** and enter **Code**: **361120***

Tangent

The tangent (tan) function has a period of 180° or π radians, which means that its graph makes one complete cycle every 180° or π radians. The x-axis represents the angle measure, and the y-axis represents the tangent of the angle. The graph of the tangent function is a series of smooth curves that cross the x-axis at every 180° or π radians and have an asymptote every $k \cdot 90°$ or $\frac{k\pi}{2}$ radians, where k is an odd integer. This can be explained by the fact that the tangent is calculated by dividing the sine by the cosine, since the cosine equals zero at those asymptote points.

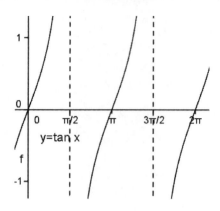

> ➢ **Review Video: Finding Tangent**
> Visit *mometrix.com/academy* and enter *Code*: **947639**

Defined and Reciprocal Functions

The tangent function is defined as the ratio of the sine to the cosine:

Tangent (tan):

$$\tan x = \frac{\sin x}{\cos x}$$

To take the reciprocal of a number means to place that number as the denominator of a fraction with a numerator of 1. The reciprocal functions are thus defined quite simply.

Cosecant (csc):

$$\csc x = \frac{1}{\sin x}$$

Secant (sec):

$$\sec x = \frac{1}{\cos x}$$

Cotangent (cot):

$$\cot x = \frac{1}{\tan x}$$

It is important to know these reciprocal functions, but they are not as commonly used as the three basic functions.

Inverse Functions

Each of the trigonometric functions accepts an angular measure, either degrees or radians, and gives a numerical value as the output. The inverse functions do the opposite; they accept a numerical value and give an angular measure as the output. The inverse sine, or arcsine, commonly written as either $\sin^{-1} x$ or $\arcsin x$, gives the angle whose sine is x. Similarly:

The inverse of $\cos x$ is written as $\cos^{-1} x$ or $\arccos x$ and means the angle whose cosine is x.
The inverse of $\tan x$ is written as $\tan^{-1} x$ or $\arctan x$ and means the angle whose tangent is x.
The inverse of $\csc x$ is written as $\csc^{-1} x$ or $\text{arccsc } x$ and means the angle whose cosecant is x.
The inverse of $\sec x$ is written as $\sec^{-1} x$ or $\text{arcsec } x$ and means the angle whose secant is x.
The inverse of $\cot x$ is written as $\cot^{-1} x$ or $\text{arccot } x$ and means the angle whose cotangent is x.

> ➤ **Review Video: <u>Inverse of a Cosine</u>**
> *Visit **mometrix.com/academy** and enter **Code**: **156054**

> ➤ **Review Video: <u>Inverse of a Tangent</u>**
> *Visit **mometrix.com/academy** and enter **Code**: **229055**

<u>Important note about solving trigonometric equations</u>
Trigonometric and algebraic equations are solved following the same rules, but while algebraic expressions have one unique solution, trigonometric equations could have multiple solutions, and you must find them all. When solving for an angle with a known trigonometric value, you must consider the sign and include all angles with that value. Your calculator will probably only give one value as an answer, typically in the following ranges:

For the inverse sine function, $\left[-\frac{\pi}{2}, \frac{\pi}{2}\right]$ or $[-90°, 90°]$
For the inverse cosine function, $[0, \pi]$ or $[0°, 180°]$
For the inverse tangent function, $\left[-\frac{\pi}{2}, \frac{\pi}{2}\right]$ or $[-90°, 90°]$

It is important to determine if there is another angle in a different quadrant that also satisfies the problem. To do this, find the other quadrant(s) with the same sign for that trigonometric function and find the angle that has the same reference angle. Then check whether this angle is also a solution.

In the first quadrant, all six trigonometric functions are positive (sin, cos, tan, csc, sec, cot).
In the second quadrant, sin and csc are positive.
In the third quadrant, tan and cot are positive.
In the fourth quadrant, cos and sec are positive.

If you remember the phrase, "ALL Students Take Classes," you will be able to remember the sign of each trigonometric function in each quadrant. ALL represents all the signs in the first quadrant. The "S" in "Students" represents the sine function and its reciprocal in the second quadrant. The "T" in "Take" represents the tangent function and its reciprocal in the third quadrant. The "C" in "Classes" represents the cosine function and its reciprocal.

Trigonometric Identities
<u>Sum and Difference</u>
To find the sine, cosine, or tangent of the sum or difference of two angles, use one of the following formulas:

$$\sin(\alpha \pm \beta) = \sin \alpha \cos \beta \pm \cos \alpha \sin \beta$$
$$\cos(\alpha \pm \beta) = \cos \alpha \cos \beta \mp \sin \alpha \sin \beta$$
$$\tan(\alpha \pm \beta) = \frac{\tan \alpha \pm \tan \beta}{1 \mp \tan \alpha \tan \beta}$$

where α and β are two angles with known sine, cosine, or tangent values as needed.

<u>Half angle</u>
To find the sine or cosine of half of a known angle, use the following formulas:

$$\sin\frac{\theta}{2} = \pm\sqrt{\frac{1-\cos\theta}{2}}$$

$$\cos\frac{\theta}{2} = \pm\sqrt{\frac{1+\cos\theta}{2}}$$

where θ is an angle with a known exact cosine value.

To determine the sign of the answer, you must notice the quadrant the given angle is in and apply the correct sign for the trigonometric function you are using. If you need to find the exact sine or cosine of an angle that you do not know, such as sin 22.5°, you can rewrite the given angle as a half angle, such as $\sin\frac{45°}{2}$, and use the formula above.

To find the tangent or cotangent of half of a known angle, use the following formulas:

$$\tan\frac{\theta}{2} = \frac{\sin\theta}{1+\cos\theta}$$
$$\cot\frac{\theta}{2} = \frac{\sin\theta}{1-\cos\theta}$$

where θ is an angle with known exact sine and cosine values.

These formulas will work for finding the tangent or cotangent of half of any angle unless the cosine of θ happens to make the denominator of the identity equal to 0.

<u>Double angles</u>
In each case, use one of the Double Angle Formulas. To find the sine or cosine of twice a known angle, use one of the following formulas:

$$\sin(2\theta) = 2\sin\theta\cos\theta$$
$$\cos(2\theta) = \cos^2\theta - \sin^2\theta \ \text{ or}$$
$$\cos(2\theta) = 2\cos^2\theta - 1 \ \text{ or}$$
$$\cos(2\theta) = 1 - 2\sin^2\theta$$

To find the tangent or cotangent of twice a known angle, use the formulas:

$$\tan(2\theta) = \frac{2\tan\theta}{1-\tan^2\theta}$$
$$\cot(2\theta) = \frac{\cot\theta - \tan\theta}{2}$$

In each case, θ is an angle with known exact sine, cosine, tangent, and cotangent values.

Products

To find the product of the sines and cosines of two different angles, use one of the following formulas:

$$\sin \alpha \sin \beta = \frac{1}{2}[\cos(\alpha - \beta) - \cos(\alpha + \beta)]$$
$$\cos \alpha \cos \beta = \frac{1}{2}[\cos(\alpha + \beta) + \cos(\alpha - \beta)]$$
$$\sin \alpha \cos \beta = \frac{1}{2}[\sin(\alpha + \beta) + \sin(\alpha - \beta)]$$
$$\cos \alpha \sin \beta = \frac{1}{2}[\sin(\alpha + \beta) - \sin(\alpha - \beta)]$$

where α and β are two unique angles.

Complementary

The trigonometric cofunction identities use the trigonometric relationships of complementary angles (angles whose sum is 90°). These are:

$$\cos x = \sin(90° - x)$$
$$\csc x = \sec(90° - x)$$
$$\cot x = \tan(90° - x)$$

Pythagorean Theorem

The Pythagorean Theorem states that $a^2 + b^2 = c^2$ for all right triangles. The trigonometric identity that derives from this principles is stated in this way:

$$\sin^2 \theta + \cos^2 \theta = 1$$

Dividing each term by either $\sin^2 \theta$ or $\cos^2 \theta$ yields two other identities, respectively:

$$1 + \cot^2 \theta = \csc^2 \theta$$
$$\tan^2 \theta + 1 = \sec^2 \theta$$

Unit Circle

A unit circle is a circle with a radius of 1 that has its center at the origin. The equation of the unit circle is $x^2 + y^2 = 1$. Notice that this is an abbreviated version of the standard equation of a circle. Because the center is the point $(0, 0)$, the values of h and k in the general equation are equal to zero and the equation simplifies to this form.

Standard Position is the position of an angle of measure θ whose vertex is at the origin, the initial side crosses the unit circle at the point $(1, 0)$, and the terminal side crosses the unit circle at some other point (a, b). In the standard position, $\sin \theta = b$, $\cos \theta = a$, and $\tan \theta = \frac{b}{a}$.

> ➢ **Review Video: <u>Unit Circles and Standard Position</u>**
> *Visit **mometrix.com/academy** and enter **Code**: 333922*

Rectangular coordinates are those that lie on the square grids of the Cartesian plane. They should be quite familiar to you. The polar coordinate system is based on a circular graph, rather than the square grid of the Cartesian system. Points in the polar coordinate system are in the format (r, θ), where r is the distance from the origin (think radius of the circle) and θ is the smallest positive angle (moving counterclockwise around the circle) made with the positive horizontal axis.

> ➢ **Review Video:** Rectangular and Polar Coordinate System
> *Visit **mometrix.com/academy** and enter **Code**: 694585*

To convert a point from rectangular (x, y) format to polar (r, θ) format, use the formula
(x, y) to $(r, \theta) \Rightarrow r = \sqrt{x^2 + y^2}$; $\theta = \arctan \frac{y}{x}$ when $x \neq 0$

> ➢ **Review Video:** Converting Between Polar and Rectangular Formats
> *Visit **mometrix.com/academy** and enter **Code**: 281325*

If x is positive, use the positive square root value for r. If x is negative, use the negative square root value for r.
If x = 0, use the following rules:
If x = 0 and y = 0, then $\theta = 0$
If x = 0 and y > 0, then $\theta = \frac{\pi}{2}$
If x = 0 and y < 0, then $\theta = \frac{3\pi}{2}$

To convert a point from polar (r, θ) format to rectangular (x, y) format, use the formula
(r, θ) to $(x, y) \Rightarrow x = r \cos \theta$; $y = r \sin \theta$

Table of commonly encountered angles

$0° = 0$ radians, $30° = \frac{\pi}{6}$ radians, $45° = \frac{\pi}{4}$ radians, $60° = \frac{\pi}{3}$ radians, and $90° = \frac{\pi}{2}$ radians

$\sin 0° = 0$	$\cos 0° = 1$	$\tan 0° = 0$
$\sin 30° = \frac{1}{2}$	$\cos 30° = \frac{\sqrt{3}}{2}$	$\tan 30° = \frac{\sqrt{3}}{3}$
$\sin 45° = \frac{\sqrt{2}}{2}$	$\cos 45° = \frac{\sqrt{2}}{2}$	$\tan 45° = 1$
$\sin 60° = \frac{\sqrt{3}}{2}$	$\cos 60° = \frac{1}{2}$	$\tan 60° = \sqrt{3}$
$\sin 90° = 1$	$\cos 90° = 0$	$\tan 90° = $ undefined
$\csc 0° = $ undefined	$\sec 0° = 1$	$\cot 0° = $ undefined
$\csc 30° = 2$	$\sec 30° = \frac{2\sqrt{3}}{3}$	$\cot 30° = \sqrt{3}$
$\csc 45° = \sqrt{2}$	$\sec 45° = \sqrt{2}$	$\cot 45° = 1$
$\csc 60° = \frac{2\sqrt{3}}{3}$	$\sec 60° = 2$	$\cot 60° = \frac{\sqrt{3}}{3}$
$\csc 90° = 1$	$\sec 90° = $ undefined	$\cot 90° = 0$

The values in the upper half of this table are values you should have memorized or be able to find quickly.

Copyright © Mometrix Media. You have been licensed one copy of this document for personal use only. Any other reproduction or redistribution is strictly prohibited. All rights reserved.

Calculus

Limits

The limit of a function is represented by the notation $\lim_{x \to a} f(x)$. It is read as "the limit of f of x as x approaches a." In many cases, $\lim_{x \to a} f(x)$ will simply be equal to $f(a)$, but not always. Limits are important because some functions are not defined or are not easy to evaluate at certain values of x.

The limit at the point is said to exist only if the limit is the same when approached from the right side as from the left: $\lim_{x \to a^+} f(x) = \lim_{x \to a^-} f(x)$). Notice the symbol by the a in each case. When x approaches a from the right, it approaches from the positive end of the number line. When x approaches a from the left, it approaches from the negative end of the number line.

If the limit as x approaches a differs depending on the direction from which it approaches, then the limit does not exist at a. In other words, if $\lim_{x \to a^+} f(x)$ does not equal $\lim_{x \to a^-} f(x)$, then the limit does not exist at a. The limit also does not exist if either of the one-sided limits does not exist.

Situations in which the limit does not exist include a function that jumps from one value to another at a, one that oscillates between two different values as x approaches a, or one that increases or decreases without bounds as x approaches a. If the limit you calculate has a value of $\frac{c}{0}$, where c is any constant, this means the function goes to infinity and the limit does not exist.

It is possible for two functions that do not have limits to be multiplied to get a new function that does have a limit. Just because two functions do not have limits, do not assume that the product will not have a limit.

The first thing to try when looking for a limit is direct substitution. To find the limit of a function $\lim_{x \to a} f(x)$ by direct substitution, substitute the value of a for x in the function and solve. The following patterns apply to finding the limit of a function by direct substitution:

$\lim_{x \to a} b = b$, where b is any real number

$\lim_{x \to a} x = a$

$\lim_{x \to a} x^n = a^n$, where n is any positive integer

$\lim_{x \to a} \sqrt{x} = \sqrt{a}; a > 0$

$\lim_{x \to a} \sqrt[n]{x} = \sqrt[n]{a}$, where n is a positive integer and $a > 0$ for all even values of n

$\lim_{x \to a} \frac{1}{x} = \frac{1}{a}; a \neq 0$

You can also use substitution for finding the limit of a trigonometric function, a polynomial function, or a rational function. Be sure that in manipulating an expression to find a limit that you do not divide by terms equal to zero.

In finding the limit of a composite function, begin by finding the limit of the innermost function. For example, to find $\lim_{x \to a} f(g(x))$, first find the value of $\lim_{x \to a} g(x)$. Then substitute this value for x in $f(x)$ and solve. The result is the limit of the original problem.

Sometimes solving $\lim_{x \to a} \frac{f(x)}{g(x)}$ by the direct substitution method will result in the numerator and denominator both being equal to zero, or both being equal to infinity. This outcome is called an indeterminate form. The limit cannot be directly found by substitution in these cases. L'Hôpital's rule is a useful method for finding the limit of a problem in the indeterminate form. L'Hôpital's rule allows you to find the limit using derivatives. Assuming both the numerator and denominator are differentiable, and that both are equal to zero when the direct substitution method is used, take the derivative of both the numerator and the denominator and then use the direct substitution method. For example, if $\lim_{x \to a} \frac{f(x)}{g(x)} = \frac{0}{0}$, take the derivatives of $f(x)$ and $g(x)$ and then find $\lim_{x \to a} \frac{f'(x)}{g'(x)}$. If $g'(x) \neq 0$, then you have found the limit of the original function. If $g'(x) = 0$ and $f'(x) = 0$, L'Hôpital's rule may be applied to the function $\frac{f'(x)}{g'(x)}$, and so on until either a limit is found, or it can be determined that the limit does not exist.

When finding the limit of the sum or difference of two functions, find the limit of each individual function and then add or subtract the results. For example, $\lim_{x \to a}[f(x) \pm g(x)] = \lim_{x \to a} f(x) \pm \lim_{x \to a} g(x)$.

To find the limit of the product or quotient of two functions, find the limit of each individual function and the multiply or divide the results. For example, $\lim_{x \to a}[f(x) \cdot g(x)] = \lim_{x \to a} f(x) \cdot \lim_{x \to a} g(x)$ and $\lim_{x \to a} \frac{f(x)}{g(x)} = \frac{\lim_{x \to a} f(x)}{\lim_{x \to a} g(x)}$, where $g(x) \neq 0$ and $\lim_{x \to a} g(x) \neq 0$. When finding the quotient of the limits of two functions, make sure the denominator is not equal to zero. If it is, use differentiation or L'Hôpital's rule to find the limit.

To find the limit of a power of a function or a root of a function, find the limit of the function and then raise the limit to the original power or take the root of the limit. For example, $\lim_{x \to a}[f(x)]^n = [\lim_{x \to a} f(x)]^n$ and $\lim_{x \to a} \sqrt[n]{f(x)} = \sqrt[n]{\lim_{x \to a} f(x)}$, where n is a positive integer and $\lim_{x \to a} f(x) > 0$ for all even values of n.
To find the limit of a function multiplied by a scalar, find the limit of the function and multiply the result by the scalar. For example, $\lim_{x \to a} kf(x) = k \lim_{x \to a} f(x)$, where k is a real number.

The squeeze theorem is known by many names, including the sandwich theorem, the sandwich rule, the squeeze lemma, the squeezing theorem, and the pinching theorem. No matter what you call it, the principle is the same. To prove the limit of a difficult function exists, find the limits of two functions, one on either side of the unknown, that are easy to compute. If the limits of these functions are equal, then that is also the limit of the unknown function. In mathematical terms, the theorem is:
If $g(x) \leq f(x) \leq h(x)$ for all values of x where $f(x)$ is the function with the unknown limit, and if $\lim_{x \to a} g(x) = \lim_{x \to a} h(x)$, then this limit is also equal to $\lim_{x \to a} f(x)$.

To find the limit of an expression containing an absolute value sign, take the absolute value of the limit. If $\lim_{n \to \infty} a_n = L$, where L is the numerical value for the limit, then $\lim_{n \to \infty} |a_n| = |L|$. Also, if $\lim_{n \to \infty} |a_n| = 0$, then $\lim_{n \to \infty} a_n = 0$. The trick comes when you are asked to find the limit as n approaches from the left. Whenever the limit is being approached from the left, it is being approached from the negative end of the domain. The absolute value sign makes everything in the equation positive, essentially eliminating the negative side of the domain. In this case, rewrite the equation without the absolute value signs and add a negative sign in front of the expression. For example, $\lim_{n \to 0^-} |x|$ becomes $\lim_{n \to 0^-}(-x)$.

Derivatives

The derivative of a function is a measure of how much that function is changing at a specific point, and is the slope of a line tangent to a curve at the specific point. The derivative of a function $f(x)$ is written $f'(x)$, and read, "f prime of x." Other notations for the derivative include $D_x f(x)$, y', $D_x y$, $\frac{dy}{dx}$, and $\frac{d}{dx} f(x)$. The definition of the derivative of a function is $f'(x) = \lim_{h \to 0} \frac{f(x+h) - f(x)}{h}$. However, this formula is rarely used.

There is a simpler method you can use to find the derivative of a polynomial. Given a function $f(x) = a_n x^n + a_{n-1} x^{n-1} + a_{n-2} x^{n-2} + \cdots + a_1 x + a_0$, multiply each exponent by its corresponding coefficient to get the new coefficient and reduce the value of the exponent by one. Coefficients with no variable are dropped. This gives $f'(x) = n a_n x^{n-1} + (n-1) a_{n-1} x^{n-2} + \cdots + a_1$, a pattern that can be repeated for each successive derivative.

Differentiable functions are functions that have a derivative. Some basic rules for finding derivatives of functions are

$f(x) = c \Rightarrow f'(x) = 0$; where c is a constant
$f(x) = x \Rightarrow f'(x) = 1$
$(cf(x))' = cf'(x)$; where c is a constant
$f(x) = x^n \Rightarrow f'(x) = nx^{n-1}$; where n is a real number
$(f + g)'(x) = f'(x) + g'(x)$
$(fg)'(x) = f(x)g'(x) + f'(x)g(x)$
$\left(\frac{f}{g}\right)'(x) = \frac{f'(x)g(x) - f(x)g'(x)}{[g(x)]^2}$
$(f \circ g)'(x) = f'(g(x)) \cdot g'(x)$

This last formula is also known as the Chain Rule. If you are finding the derivative of a polynomial that is raised to a power, let the polynomial be represented by $g(x)$ and use the Chain Rule. The chain rule is one of the most important concepts to grasp in the early stages of learning calculus. Many other rules and shortcuts are based upon the chain rule.

These rules may also be used to take multiple derivatives of the same function. The derivative of the derivative is called the second derivative and is represented by the notation $f''(x)$. Taking one more derivative, if possible, gives the third derivative and is represented by the notation $f'''(x)$ or $f^{(3)}(x)$.

Other derivative functions

An implicit function is one where it is impossible, or very difficult, to express one variable in terms of another by normal algebraic methods. This would include functions that have both variables raised to a power greater than 1, functions that have two variables multiplied by each other, or a combination of the two. To differentiate such a function with respect to x, take the derivate of each term that contains a variable, either x or y. When differentiating a term with y, use the chain rule, first taking the derivative with respect to y, and then multiplying by $\frac{dy}{dx}$. If a term contains both x

- 35 -

and y, you will have to use the product rule as well as the chain rule. Once the derivative of each individual term has been found, use the rules of algebra to solve for $\frac{dy}{dx}$ to get the final answer.

Example: Find $\frac{dy}{dx}$ given the equation $xy^2 = 3y + 2x$. Take the derivative of each term with respect to x: $y^2 + 2xy\frac{dy}{dx} = 3\frac{dy}{dx} + 2$. Note that the first term in the original equation required the use of the product rule and the chain rule. Using algebra, isolate $\frac{dy}{dx}$ on one side of the equation to yield $\frac{dy}{dx} = \frac{y^2-2}{3-2xy}$.

Trigonometric functions are any functions that include one of the six trigonometric expressions. The following rules for derivatives apply for all trigonometric differentiation:
$$\frac{d}{dx}(\sin x) = \cos x, \frac{d}{dx}(\cos x) = -\sin x, \frac{d}{dx}(\tan x) = \sec^2 x$$

For functions that are a combination of trigonometric and algebraic expressions, use the chain rule:
$$\frac{d}{dx}(\sin u) = \cos u \frac{du}{dx}$$
$$\frac{d}{dx}(\cos u) = -\sin u \frac{du}{dx}$$
$$\frac{d}{dx}(\tan u) = \sec^2 u \frac{du}{dx}.$$
$$\frac{d}{dx}(\sec u) = \tan u \sec u \frac{du}{dx}.$$
$$\frac{d}{dx}(\csc u) = -\csc u \cot u \frac{du}{dx}$$
$$\frac{d}{dx}(\cot u) = -\csc^2 u \frac{du}{dx}$$

Functions involving the inverses of the trigonometric functions can also be differentiated.
$$\frac{d}{dx}(\sin^{-1} u) = \frac{1}{\sqrt{1-u^2}}\frac{du}{dx}$$
$$\frac{d}{dx}(\cos^{-1} u) = \frac{-1}{\sqrt{1-u^2}}\frac{du}{dx}$$
$$\frac{d}{dx}(\tan^{-1} u) = \frac{1}{1+u^2}\frac{du}{dx}$$
$$\frac{d}{dx}(\csc^{-1} u) = \frac{-1}{u\sqrt{u^2-1}}\frac{du}{dx}$$
$$\frac{d}{dx}(\sec^{-1} u) = \frac{1}{u\sqrt{u^2-1}}\frac{du}{dx}$$
$$\frac{d}{dx}(\cot^{-1} u) = \frac{-1}{1+u^2}\frac{du}{dx}$$

In each of the above expressions, u represents a differentiable function. Also, the value of u must be such that the radicand, if applicable, is a positive number. Remember the expression $\frac{du}{dx}$ means to take the derivative of the function u with respect to the variable x.

Exponential functions are in the form e^x, which has itself as its derivative: $\frac{d}{dx}e^x = e^x$. For functions that have a function as the exponent rather than just an x, use the formula $\frac{d}{dx}e^u = e^u\frac{du}{dx}$.

The inverse of the exponential function is the natural logarithm. To find the derivative of the natural logarithm, use the formula $\frac{d}{dx}\ln u = \frac{1}{u}\frac{du}{dx}$.

If you are trying to solve an expression with a variable in the exponent, use the formula $a^x = e^{x \ln a}$, where a is a positive real number and x is any real number. To find the derivative of a function in this format, use the formula $\frac{d}{dx} a^x = a^x \ln a$. If the exponent is a function rather than a single variable x, use the formula $\frac{d}{dx} a^u = a^u \ln a \frac{du}{dx}$.

If you are trying to solve an expression involving a logarithm, use the formula $\frac{d}{dx}(\log_a x) = \frac{1}{x \ln a}$ or $\frac{d}{dx}(\log_a |u|) = \frac{1}{u \ln a}\frac{du}{dx}; u \neq 0$.

Characteristics of functions (using calculus)

Rolle's Theorem states that if a differentiable function has two different values in the domain that correspond to a single value in the range, then the function must have a point between them where the slope of the tangent to the graph is zero. This point will be a maximum or a minimum value of the function between those two points. The maximum or minimum point is the point at which $f'(c) = 0$, where c is within the appropriate interval of the function's domain. The following graph shows a function with one maximum in the second quadrant and one minimum in the fourth quadrant.

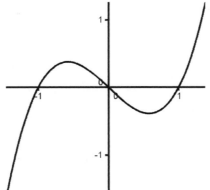

According to the Mean Value Theorem, between any two points on a curve, there exists a tangent to the curve whose slope is parallel to the chord formed by joining those two points. Remember the formula for slope: $m = \frac{\Delta x}{\Delta y}$. In a function, $f(x)$ represents the value for y. Therefore, if you have two points on a curve, m and n, the corresponding points are $(m, f(m))$ and $(n, f(n))$. Assuming $m < n$, the formula for the slope of the chord joining those two points is $\frac{f(n)-f(m)}{n-m}$. This must also be the slope of a line parallel to the chord, since parallel lines have equal slopes. Therefore, there must be a value p between m and n such that $f'(p) = \frac{f(n)-f(m)}{n-m}$.

For a function to have continuity, its graph must be an unbroken curve. That is, it is a function that can be graphed without having to lift the pencil to move it to a different point. To say a function is continuous at point p, you must show the function satisfies three requirements. First, $f(p)$ must exist. If you evaluate the function at p, it must yield a real number. Second, there must exist a relationship such that $\lim_{x \to p} f(x) = f(p)$. Finally, the following relationship must be true:
$$\lim_{x \to p^+} F(x) = \lim_{x \to p^-} F(x) = F(p)$$
If all three of these requirements are met, a function is considered continuous at p. If any one of them is not true, the function is not continuous at p.

Tangents are lines that touch a curve in exactly one point and have the same slope as the curve at that point. To find the slope of a curve at a given point and the slope of its tangent line at that point, find the derivative of the function of the curve. If the slope is undefined, the tangent is a vertical line. If the slope is zero, the tangent is a horizontal line.

A line that is normal to a curve at a given point is perpendicular to the tangent at that point. Assuming $f'(x) \neq 0$, the equation for the normal line at point (a, b) is: $y - b = -\frac{1}{f'(a)}(x - a)$. The easiest way to find the slope of the normal is to take the negative reciprocal of the slope of the tangent. If the slope of the tangent is zero, the slope of the normal is undefined. If the slope of the tangent is undefined, the slope of the normal is zero.

Rectilinear motion is motion along a straight line rather than a curved path. This concept is generally used in problems involving distance, velocity, and acceleration.

Average velocity over a period of time is found using the formula $\bar{v} = \frac{s(t_2) - s(t_1)}{t_2 - t_1}$, where t_1 and t_2 are specific points in time and $s(t_1)$ and $s(t_2)$ are the distances traveled at those points in time. Instantaneous velocity at a specific time is found using the formula $v = \lim_{h \to 0} \frac{s(t+h) - s(t)}{h}$, or $v = s'(t)$.

Remember that velocity at a given point is found using the first derivative, and acceleration at a given point is found using the second derivative. Therefore, the formula for acceleration at a given point in time is found using the formula $a(t) = v'(t) = s''(t)$, where a is acceleration, v is velocity, and s is distance or location.

Remember Rolle's Theorem, which stated that if two points have the same value in the range that there must be a point between them where the slope of the graph is zero. This point is located at a peak or valley on the graph. A peak is a maximum point, and a valley is a minimum point. The relative minimum is the lowest point on a graph for a given section of the graph. It may or may not be the same as the absolute minimum, which is the lowest point on the entire graph. The relative maximum is the highest point on one section of the graph. Again, it may or may not be the same as the absolute maximum. A relative extremum (plural extrema) is a relative minimum or relative maximum point on a graph.

A critical point is a point $(x, f(x))$ that is part of the domain of a function, such that either $f'(x) = 0$ or $f'(x)$ does not exist. If either of these conditions is true, then x is either an inflection point or a point at which the slope of the curve changes sign. If the slope changes sign, then a relative minimum or maximum occurs.

In graphing an equation with relative extrema, use a sign diagram to approximate the shape of the graph. Once you have determined the relative extrema, calculate the sign of a point on either side of each critical point. This will give a general shape of the graph, and you will know whether each critical point is a relative minimum, a relative maximum, or a point of inflection.

Remember that critical points occur where the slope of the curve is 0. Also remember that the first derivative of a function gives the slope of the curve at a particular point on the curve. Because of this property of the first derivative, the first derivative test can be used to determine if a critical point is a minimum or maximum. If $f'(x)$ is negative at a point to the left of a critical number and $f'(x)$ is positive at a point to the right of a critical number, then the critical number is a relative minimum. If $f'(x)$ is positive to the left of a critical number and $f'(x)$ is negative to the right of a critical number, then the critical number is a relative maximum. If $f'(x)$ has the same sign on both sides, then the critical number is a point of inflection.

The second derivative, designated by $f''(x)$, is helpful in determining whether the relative extrema of a function are relative maximums or relative minimums. If the second derivative at the critical point is greater than zero, the critical point is a relative minimum. If the second derivative at the critical point is less than zero, the critical point is a relative maximum. If the second derivative at the critical point is equal to zero, you must use the first derivative test to determine whether the point is a relative minimum or a relative maximum.

There are a couple of ways to determine the concavity of the graph of a function. To test a portion of the graph that contains a point with domain p, find the second derivative of the function and evaluate it for p. If $f''(p) > 0$, then the graph is concave upward at that point. If $f''(p) < 0$, then the graph is concave downward at that point.

The point of inflection on the graph of a function is the point at which the concavity changes from concave downward to concave upward or from concave upward to concave downward. The easiest way to find the points of inflection is to find the second derivative of the function and then solve the equation $f''(x) = 0$. Remember that if $f''(p) > 0$, the graph is concave upward, and if $f''(p) < 0$, the graph is concave downward. Logically, the concavity changes at the point when $f''(p) = 0$.

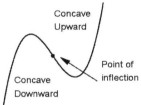

The derivative tests that have been discussed thus far can help you get a rough picture of what the graph of an unfamiliar function looks like. Begin by solving the equation $f(x) = 0$ to find all the zeros of the function, if they exist. Plot these points on the graph. Then, find the first derivative of the function and solve the equation $f'(x) = 0$ to find the critical points. Remember the numbers obtained here are the x portions of the coordinates. Substitute these values for x in the original function and solve for y to get the full coordinates of the points. Plot these points on the graph. Take the second derivative of the function and solve the equation $f''(x) = 0$ to find the points of inflection. Substitute in the original function to get the coordinates and graph these points. Test points on both sides of the critical points to test for concavity and draw the curve.

Antiderivatives (Integrals)

The antiderivative of a function is the function whose first derivative is the original function. Antiderivatives are typically represented by capital letters, while their first derivatives are represented by lower case letters. For example, if $F' = f$, then F is the antiderivative of f. Antiderivatives are also known as indefinite integrals. When taking the derivative of a function, any constant terms in the function are eliminated because their derivative is 0. To account for this possibility, when you take the indefinite integral of a function, you must add an unknown constant C

to the end of the function. Because there is no way to know what the value of the original constant was when looking just at the first derivative, the integral is indefinite.

To find the indefinite integral, reverse the process of differentiation. You will likely only need to know the most basic of integrals. Below are the formulas for constants and powers of x.

$\int 0 \, dx = C$

$\int k \, dx = kx + C$

$\int x^n \, dx = \frac{x^{n+1}}{n+1} + C$, where $n \neq -1$

Recall that in the differentiation of powers of x, you multiplied the coefficient of the term by the exponent of the variable and then reduced the exponent by one. In integration, the process is reversed: add one to the value of the exponent, and then divide the coefficient of the term by this number to get the integral. Because you do not know the value of any constant term that might have been in the original function, add C to the end of the function once you have completed this process for each term.

Finding the integral of a function is the opposite of finding the derivative of the function. Where possible, you can use the trigonometric or logarithmic differentiation formulas in reverse, and add C to the end to compensate for the unknown term. In instances where a negative sign appears in the differentiation formula, move the negative sign to the opposite side (multiply both sides by -1) to reverse for the integration formula. You should end up with the following formulas:

$\int \cos x \, dx = \sin x + C$

$\int \sec x \tan x \, dx = \sec x + C$

$\int \sin x \, dx = -\cos x + C$

$\int \csc x \cot x \, dx = -\csc x + C$

$\int \sec^2 x \, dx = \tan x + C$

$\int \csc^2 x \, dx = -\cot x + C$

$\int \frac{1}{x} \, dx = \ln |x| + C$

$\int e^x \, dx = e^x + C$

Integration by substitution is the integration version of the chain rule for differentiation. The formula for integration by substitution is given by the equation

$$f(g(x))g'(x)dx = \int f(u)du \, ; u = g(x) \text{ and } du = g'(x)dx.$$

When a function is in a format that is difficult or impossible to integrate using traditional integration methods and formulas due to multiple functions being combined, use the formula shown above to convert the function to a simpler format that can be integrated directly.

Integration by parts is the integration version of the product rule for differentiation. Whenever you are asked to find the integral of the product of two different functions or parts, integration by parts can make the process simpler. Recall for differentiation $(fg)'(x) = f(x)g'(x) + g(x)f'(x)$. This can also be written $\frac{d}{dx}(u \cdot v) = u\frac{dv}{dx} + v\frac{du}{dx}$, where $u = f(x)$ and $v = g(x)$. Rearranging to integral form gives the formula

$$\int u \, dv = uv - \int v \, du$$

which can also be written as

$$\int f(x)g'(x) \, dx = f(x)g(x) - \int f'(x)g(x) \, dx$$

When using integration by parts, the key is selecting the best functions to substitute for u and v so that you make the integral easier to solve and not harder.

While the indefinite integral has an undefined constant added at the end, the definite integral can be calculated as an exact real number. To find the definite integral of a function over a closed interval, use the formula

$$\int_n^m f(x) \, dx = F(m) - F(n)$$

where F is the integral of f. Because you have been given the boundaries of n and m, no undefined constant C is needed.

The First Fundamental Theorem of Calculus shows that the process of indefinite integration can be reversed by finding the first derivative of the resulting function. It also gives the relationship between differentiation and integration over a closed interval of the function. For example, assuming a function is continuous over the interval $[m, n]$, you can find the definite integral by using the formula $\int_m^n f(x) \, dx = F(n) - F(m)$. Many times the notation $\int_m^n f(x) \, dx = F(x) \big|_m^n = F(n) - F(m)$ is also used to represent the Fundamental Theorem of Calculus. To find the average value of the function over the given interval, use the formula $\frac{1}{n-m} \int_m^n f(x) \, dx$.

The Second Fundamental Theorem of Calculus is related to the first. This theorem states that, assuming the function is continuous over the interval you are considering, taking the derivative of the integral of a function will yield the original function. The general format for this theorem is $\frac{d}{dx} \int_c^x f(x) \, dx = f(x)$ for any point having a domain value equal to c in the given interval.

For each of the following properties of integrals of function f, the variables m, n, and p represent values in the domain of the given interval of $f(x)$. The function is assumed to be integrable across all relevant intervals.

$$\int_n^n f(x) \, dx = 0$$

$$\int_m^n f(x) \, dx = -\int_n^m f(x) \, dx$$

$$\int_m^n kf(x) \, dx = k\int_m^n f(x) \, dx$$

$$\int_m^n f(x) \, dx = \int_m^p f(x) \, dx + \int_p^n f(x) \, dx$$

If $f(x)$ is an even function, then

$$\int_{-m}^m f(x) \, dx = 2\int_0^m f(x) \, dx$$

If $f(x)$ is an odd function, then

$$\int_{-m}^{m} f(x)\, dx = 0$$

The trapezoidal rule, also called the trapezoid rule or the trapezium rule, is another way to approximate the area under a curve. In this case, a trapezoid is drawn such that one side of the trapezoid is on the x-axis and the parallel sides terminate on points of the curve so that part of the curve is above the trapezoid and part of the curve is below the trapezoid. While this does not give an exact area for the region under the curve, it does provide a close approximation.

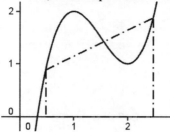

Geometry

Lines and Planes

A point is a fixed location in space; has no size or dimensions; commonly represented by a dot.

A line is a set of points that extends infinitely in two opposite directions. It has length, but no width or depth. A line can be defined by any two distinct points that it contains. A line segment is a portion of a line that has definite endpoints. A ray is a portion of a line that extends from a single point on that line in one direction along the line. It has a definite beginning, but no ending.

A plane is a two-dimensional flat surface defined by three non-collinear points. A plane extends an infinite distance in all directions in those two dimensions. It contains an infinite number of points, parallel lines and segments, intersecting lines and segments, as well as parallel or intersecting rays. A plane will never contain a three-dimensional figure or skew lines. Two given planes will either be parallel or they will intersect to form a line. A plane may intersect a circular conic surface, such as a cone, to form conic sections, such as the parabola, hyperbola, circle or ellipse.

Perpendicular lines are lines that intersect at right angles. They are represented by the symbol ⊥. The shortest distance from a line to a point not on the line is a perpendicular segment from the point to the line.

Parallel lines are lines in the same plane that have no points in common and never meet. It is possible for lines to be in different planes, have no points in common, and never meet, but they are not parallel because they are in different planes.

A bisector is a line or line segment that divides another line segment into two equal lengths. A perpendicular bisector of a line segment is composed of points that are equidistant from the endpoints of the segment it is dividing.

Intersecting lines are lines that have exactly one point in common. Concurrent lines are multiple lines that intersect at a single point.

A transversal is a line that intersects at least two other lines, which may or may not be parallel to one another. A transversal that intersects parallel lines is a common occurrence in geometry.

Angles

An angle is formed when two lines or line segments meet at a common point. It may be a common starting point for a pair of segments or rays, or it may be the intersection of lines. Angles are represented by the symbol ∠.

The vertex is the point at which two segments or rays meet to form an angle. If the angle is formed by intersecting rays, lines, and/or line segments, the vertex is the point at which four angles are formed. The pairs of angles opposite one another are called vertical angles, and their measures are equal.

An acute angle is an angle with a degree measure less than 90°.
A right angle is an angle with a degree measure of exactly 90°.
An obtuse angle is an angle with a degree measure greater than 90° but less than 180°.
A straight angle is an angle with a degree measure of exactly 180°. This is also a semicircle.
A reflex angle is an angle with a degree measure greater than 180° but less than 360°.
A full angle is an angle with a degree measure of exactly 360°.

> ➤ **Review Video: Geometric Symbols: Angles**
> *Visit **mometrix.com/academy** and enter **Code: 452738***

Two angles whose sum is exactly 90° are said to be complementary. The two angles may or may not be adjacent. In a right triangle, the two acute angles are complementary.

Two angles whose sum is exactly 180° are said to be supplementary. The two angles may or may not be adjacent. Two intersecting lines always form two pairs of supplementary angles. Adjacent supplementary angles will always form a straight line.

Two angles that have the same vertex and share a side are said to be adjacent. Vertical angles are not adjacent because they share a vertex but no common side.

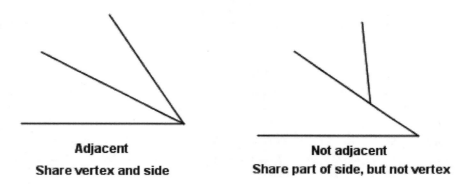

Adjacent
Share vertex and side

Not adjacent
Share part of side, but not vertex

When two parallel lines are cut by a transversal, the angles that are between the two parallel lines are interior angles. In the diagram below, angles 3, 4, 5, and 6 are interior angles.

When two parallel lines are cut by a transversal, the angles that are outside the parallel lines are exterior angles. In the diagram below, angles 1, 2, 7, and 8 are exterior angles.

When two parallel lines are cut by a transversal, the angles that are in the same position relative to the transversal and a parallel line are corresponding angles. The diagram below has four pairs of corresponding angles: angles 1 and 5; angles 2 and 6; angles 3 and 7; and angles 4 and 8. Corresponding angles formed by parallel lines are congruent.

When two parallel lines are cut by a transversal, the two interior angles that are on opposite sides of the transversal are called alternate interior angles. In the diagram below, there are two pairs of alternate interior angles: angles 3 and 6, and angles 4 and 5. Alternate interior angles formed by parallel lines are congruent.

When two parallel lines are cut by a transversal, the two exterior angles that are on opposite sides of the transversal are called alternate exterior angles.

In the diagram below, there are two pairs of alternate exterior angles: angles 1 and 8, and angles 2 and 7. Alternate exterior angles formed by parallel lines are congruent.

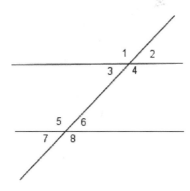

When two lines intersect, four angles are formed. The non-adjacent angles at this vertex are called vertical angles. Vertical angles are congruent. In the diagram, $\angle ABD \cong \angle CBE$ and $\angle ABC \cong \angle DBE$.

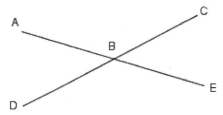

Triangles

An equilateral triangle is a triangle with three congruent sides. An equilateral triangle will also have three congruent angles, each 60°. All equilateral triangles are also acute triangles.

An isosceles triangle is a triangle with two congruent sides. An isosceles triangle will also have two congruent angles opposite the two congruent sides.

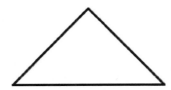

- 45 -

A scalene triangle is a triangle with no congruent sides. A scalene triangle will also have three angles of different measures. The angle with the largest measure is opposite the longest side, and the angle with the smallest measure is opposite the shortest side.

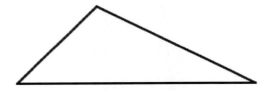

An acute triangle is a triangle whose three angles are all less than 90°. If two of the angles are equal, the acute triangle is also an isosceles triangle. If the three angles are all equal, the acute triangle is also an equilateral triangle.

A right triangle is a triangle with exactly one angle equal to 90°. All right triangles follow the Pythagorean Theorem. A right triangle can never be acute or obtuse.

An obtuse triangle is a triangle with exactly one angle greater than 90°. The other two angles may or may not be equal. If the two remaining angles are equal, the obtuse triangle is also an isosceles triangle.

Terminology

Altitude of a Triangle: A line segment drawn from one vertex perpendicular to the opposite side. In the diagram below, $\overline{BE}$, $\overline{AD}$, and $\overline{CF}$ are altitudes. The three altitudes in a triangle are always concurrent.

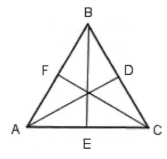

Height of a Triangle: The length of the altitude, although the two terms are often used interchangeably.

Orthocenter of a Triangle: The point of concurrency of the altitudes of a triangle. Note that in an obtuse triangle, the orthocenter will be outside the circle, and in a right triangle, the orthocenter is the vertex of the right angle.

Median of a Triangle: A line segment drawn from one vertex to the midpoint of the opposite side. This is not the same as the altitude, except the altitude to the base of an isosceles triangle and all three altitudes of an equilateral triangle.

Centroid of a Triangle: The point of concurrency of the medians of a triangle. This is the same point as the orthocenter only in an equilateral triangle. Unlike the orthocenter, the centroid is always

- 46 -

inside the triangle. The centroid can also be considered the exact center of the triangle. Any shape triangle can be perfectly balanced on a tip placed at the centroid. The centroid is also the point that is two-thirds the distance from the vertex to the opposite side.

Pythagorean Theorem

The side of a triangle opposite the right angle is called the hypotenuse. The other two sides are called the legs. The Pythagorean Theorem states a relationship among the legs and hypotenuse of a right triangle: $a^2 + b^2 = c^2$, where a and b are the lengths of the legs of a right triangle, and c is the length of the hypotenuse. Note that this formula will only work with right triangles.

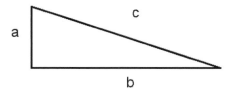

General Rules

The Triangle Inequality Theorem states that the sum of the measures of any two sides of a triangle is always greater than the measure of the third side. If the sum of the measures of two sides were equal to the third side, a triangle would be impossible because the two sides would lie flat across the third side and there would be no vertex. If the sum of the measures of two of the sides was less than the third side, a closed figure would be impossible because the two shortest sides would never meet.

The sum of the measures of the interior angles of a triangle is always 180°. Therefore, a triangle can never have more than one angle greater than or equal to 90°.

In any triangle, the angles opposite congruent sides are congruent, and the sides opposite congruent angles are congruent. The largest angle is always opposite the longest side, and the smallest angle is always opposite the shortest side.

The line segment that joins the midpoints of any two sides of a triangle is always parallel to the third side and exactly half the length of the third side.

Similarity and Congruence Rules

Similar triangles are triangles whose corresponding angles are equal and whose corresponding sides are proportional. Represented by AA. Similar triangles whose corresponding sides are congruent are also congruent triangles.

> ➤ **Review Video: Similar Triangles**
> *Visit **mometrix.com/academy** and enter* **Code: 398538**

Three sides of one triangle are congruent to the three corresponding sides of the second triangle. Represented as SSS.

Two sides and the included angle (the angle formed by those two sides) of one triangle are congruent to the corresponding two sides and included angle of the second triangle. Represented by SAS.

Two angles and the included side (the side that joins the two angles) of one triangle are congruent to the corresponding two angles and included side of the second triangle. Represented by ASA.

Two angles and a non-included side of one triangle are congruent to the corresponding two angles and non-included side of the second triangle. Represented by AAS.

Note that AAA is not a form for congruent triangles. This would say that the three angles are congruent, but says nothing about the sides. This meets the requirements for similar triangles, but not congruent triangles.

Area and Perimeter Formulas

The perimeter of any triangle is found by summing the three side lengths; $P = a + b + c$. For an equilateral triangle, this is the same as $P = 3s$, where s is any side length, since all three sides are the same length.

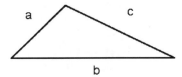

The area of any triangle can be found by taking half the product of one side length (base or b) and the perpendicular distance from that side to the opposite vertex (height or h). In equation form, $A = \frac{1}{2}bh$. For many triangles, it may be difficult to calculate h, so using one of the other formulas given here may be easier.

Another formula that works for any triangle is $A = \sqrt{s(s-a)(s-b)(s-c)}$, where A is the area, s is the semiperimeter $s = \frac{a+b+c}{2}$, and a, b, and c are the lengths of the three sides.

The area of an equilateral triangle can found by the formula $A = \frac{\sqrt{3}}{4}s^2$, where A is the area and s is the length of a side. You could use the $30° - 60° - 90°$ ratios to find the height of the triangle and then use the standard triangle area formula, but this is faster.

The area of an isosceles triangle can found by the formula, $A = \frac{1}{2}b\sqrt{a^2 - \frac{b^2}{4}}$, where A is the area, b is the base (the unique side), and a is the length of one of the two congruent sides. If you do not remember this formula, you can use the Pythagorean Theorem to find the height so you can use the standard formula for the area of a triangle.

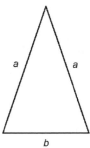

> **Review Video: <u>Area and Perimeter of a Triangle</u>**
> *Visit mometrix.com/academy and enter Code: 853779*

Trigonometric Formulas

In the diagram below, angle C is the right angle, and side c is the hypotenuse. Side a is the side adjacent to angle B and side b is the side adjacent to angle A. These formulas will work for any acute angle in a right triangle. They will NOT work for any triangle that is not a right triangle. Also, they will not work for the right angle in a right triangle, since there are not distinct adjacent and opposite sides to differentiate from the hypotenuse.

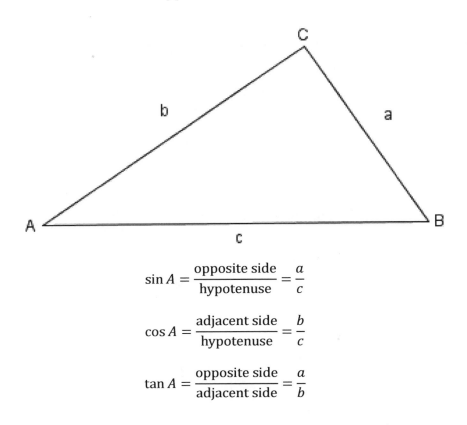

$$\sin A = \frac{\text{opposite side}}{\text{hypotenuse}} = \frac{a}{c}$$

$$\cos A = \frac{\text{adjacent side}}{\text{hypotenuse}} = \frac{b}{c}$$

$$\tan A = \frac{\text{opposite side}}{\text{adjacent side}} = \frac{a}{b}$$

- 49 -

$$\csc A = \frac{1}{\sin A} = \frac{\text{hypotenuse}}{\text{opposite side}} = \frac{c}{a}$$

$$\sec A = \frac{1}{\cos A} = \frac{\text{hypotenuse}}{\text{adjacent side}} = \frac{c}{b}$$

$$\cot A = \frac{1}{\tan A} = \frac{\text{adjacent side}}{\text{opposite side}} = \frac{b}{a}$$

Laws of Sines and Cosines

The Law of Sines states that $\frac{\sin A}{a} = \frac{\sin B}{b} = \frac{\sin C}{c}$, where A, B, and C are the angles of a triangle, and a, b, and c are the sides opposite their respective angles. This formula will work with all triangles, not just right triangles.

The Law of Cosines is given by the formula $c^2 = a^2 + b^2 - 2ab(\cos C)$, where a, b, and c are the sides of a triangle, and C is the angle opposite side c. This formula is similar to the Pythagorean Theorem, but unlike the Pythagorean Theorem, it can be used on any triangle.

> ➤ **Review Video:** <u>Cosine</u>
> Visit ***mometrix.com/academy*** *and enter* ***Code***: **361120**

Polygons

Each straight line segment of a polygon is called a side.

The point at which two sides of a polygon intersect is called the vertex. In a polygon, the number of sides is always equal to the number of vertices.

A polygon with all sides congruent and all angles equal is called a regular polygon.

A line segment from the center of a polygon perpendicular to a side of the polygon is called the apothem. In a regular polygon, the apothem can be used to find the area of the polygon using the formula $A = \frac{1}{2}ap$, where a is the apothem and p is the perimeter.

A line segment from the center of a polygon to a vertex of the polygon is called a radius. The radius of a regular polygon is also the radius of a circle that can be circumscribed about the polygon.

Triangle – 3 sides
Quadrilateral – 4 sides
Pentagon – 5 sides
Hexagon – 6 sides
Heptagon – 7 sides
Octagon – 8 sides
Nonagon – 9 sides
Decagon – 10 sides
Dodecagon – 12 sides

More generally, an *n*-gon is a polygon that has *n* angles and *n* sides.

The sum of the interior angles of an *n*-sided polygon is $(n-2)180°$. For example, in a triangle n = 3, so the sum of the interior angles is $(3-2)180° = 180°$. In a quadrilateral, n = 4, and the sum of the angles is $(4-2)180° = 360°$. The sum of the interior angles of a polygon is equal to the sum of the interior angles of any other polygon with the same number of sides.

A diagonal is a line segment that joins two non-adjacent vertices of a polygon.
A convex polygon is a polygon whose diagonals all lie within the interior of the polygon.
A concave polygon is a polygon with a least one diagonal that lies outside the polygon. In the diagram below, quadrilateral *ABCD* is concave because diagonal $\overline{AC}$ lies outside the polygon.

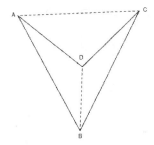

The number of diagonals a polygon has can be found by using the formula: number of diagonals $= \frac{n(n-3)}{2}$, where *n* is the number of sides in the polygon. This formula works for all polygons, not just regular polygons.

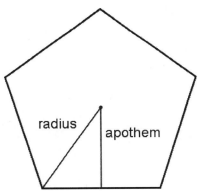

Congruent figures are geometric figures that have the same size and shape. All corresponding angles are equal, and all corresponding sides are equal. It is indicated by the symbol ≅.

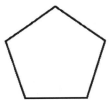 

Congruent polygons

Similar figures are geometric figures that have the same shape, but do not necessarily have the same size. All corresponding angles are equal, and all corresponding sides are proportional, but they do not have to be equal. It is indicated by the symbol ~.

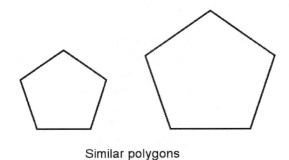

Similar polygons

Note that all congruent figures are also similar, but not all similar figures are congruent.

> **Review Video: Polygons, Similarity, and Congruence**
> Visit *mometrix.com/academy* and enter *Code*: **686174**

Line of Symmetry: The line that divides a figure or object into two symmetric parts. Each symmetric half is congruent to the other. An object may have no lines of symmetry, one line of symmetry, or more than one line of symmetry.

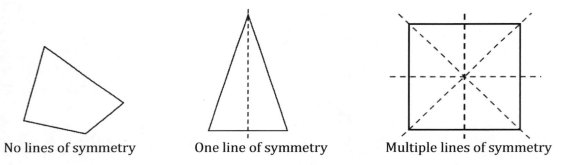

No lines of symmetry   One line of symmetry   Multiple lines of symmetry

Quadrilateral: A closed two-dimensional geometric figure composed of exactly four straight sides. The sum of the interior angles of any quadrilateral is 360°.

Parallelogram: A quadrilateral that has exactly two pairs of opposite parallel sides. The sides that are parallel are also congruent. The opposite interior angles are always congruent, and the consecutive interior angles are supplementary. The diagonals of a parallelogram bisect each other. Each diagonal divides the parallelogram into two congruent triangles.

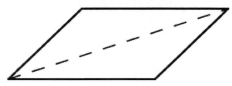

Trapezoid: Traditionally, a quadrilateral that has exactly one pair of parallel sides. Some math texts define trapezoid as a quadrilateral that has at least one pair of parallel sides. Because there are no rules governing the second pair of sides, there are no rules that apply to the properties of the diagonals of a trapezoid.

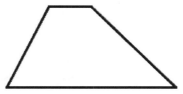

Rectangles, rhombuses, and squares are all special forms of parallelograms.

Rectangle: A parallelogram with four right angles. All rectangles are parallelograms, but not all parallelograms are rectangles. The diagonals of a rectangle are congruent.

Rhombus: A parallelogram with four congruent sides. All rhombuses are parallelograms, but not all parallelograms are rhombuses. The diagonals of a rhombus are perpendicular to each other.

> **Review Video:** <u>Diagonals of Parallelograms, Rectangles, and Rhombi</u>
> Visit **mometrix.com/academy** and enter **Code: 320040**

Square: A parallelogram with four right angles and four congruent sides. All squares are also parallelograms, rhombuses, and rectangles. The diagonals of a square are congruent and perpendicular to each other.

A quadrilateral whose diagonals bisect each other is a parallelogram. A quadrilateral whose opposite sides are parallel (2 pairs of parallel sides) is a parallelogram.

A quadrilateral whose diagonals are perpendicular bisectors of each other is a rhombus. A quadrilateral whose opposite sides (both pairs) are parallel and congruent is a rhombus.

A parallelogram that has a right angle is a rectangle. (Consecutive angles of a parallelogram are supplementary. Therefore if there is one right angle in a parallelogram, there are four right angles in that parallelogram.)

A rhombus with one right angle is a square. Because the rhombus is a special form of a parallelogram, the rules about the angles of a parallelogram also apply to the rhombus.

Area and Perimeter Formulas

The area of a square is found by using the formula $A = s^2$, where and s is the length of one side.

The perimeter of a square is found by using the formula $P = 4s$, where s is the length of one side. Because all four sides are equal in a square, it is faster to multiply the length of one side by 4 than to add the same number four times. You could use the formulas for rectangles and get the same answer.

> ➤ **Review Video:** <u>**Area and Perimeter of a Square**</u>
> Visit ***mometrix.com/academy*** and enter ***Code***: **620902**

The area of a rectangle is found by the formula $A = lw$, where A is the area of the rectangle, l is the length (usually considered to be the longer side) and w is the width (usually considered to be the shorter side). The numbers for l and w are interchangeable.

The perimeter of a rectangle is found by the formula $P = 2l + 2w$ or $P = 2(l + w)$, where l is the length, and w is the width. It may be easier to add the length and width first and then double the result, as in the second formula.

> ➤ **Review Video:** <u>**Area and Perimeter of a Rectangle**</u>
> Visit ***mometrix.com/academy*** and enter ***Code***: **933707**

The area of a parallelogram is found by the formula $A = bh$, where b is the length of the base, and h is the height. Note that the base and height correspond to the length and width in a rectangle, so this formula would apply to rectangles as well. Do not confuse the height of a parallelogram with the length of the second side. The two are only the same measure in the case of a rectangle.

The perimeter of a parallelogram is found by the formula $P = 2a + 2b$ or $P = 2(a + b)$, where a and b are the lengths of the two sides.

> ➤ **Review Video:** <u>**Area and Perimeter of a Parallelogram**</u>
> Visit ***mometrix.com/academy*** and enter ***Code***: **718313**

The area of a trapezoid is found by the formula $A = \frac{1}{2}h(b_1 + b_2)$, where h is the height (segment joining and perpendicular to the parallel bases), and b_1 and b_2 are the two parallel sides (bases). Do not use one of the other two sides as the height unless that side is also perpendicular to the parallel bases.

The perimeter of a trapezoid is found by the formula $P = a + b_1 + c + b_2$, where a, b_1, c, and b_2 are the four sides of the trapezoid.

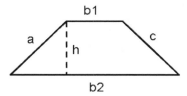

> ➤ **Review Video:** <u>Area and Perimeter of a Trapezoid</u>
> *Visit **mometrix.com/academy** and enter **Code**: 587523*

Circles

The center is the single point inside the circle that is equidistant from every point on the circle. (Point O in the diagram below.)

> ➤ **Review Video:** <u>Points of a Circle</u>
> *Visit **mometrix.com/academy** and enter **Code**: 420746*

The radius is a line segment that joins the center of the circle and any one point on the circle. All radii of a circle are equal. (Segments OX, OY, and OZ in the diagram below.)

The diameter is a line segment that passes through the center of the circle and has both endpoints on the circle. The length of the diameter is exactly twice the length of the radius. (Segment XZ in the diagram below.)

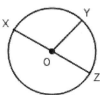

The area of a circle is found by the formula $A = \pi r^2$, where r is the length of the radius. If the diameter of the circle is given, remember to divide it in half to get the length of the radius before proceeding.

The circumference of a circle is found by the formula $C = 2\pi r$, where r is the radius. Again, remember to convert the diameter if you are given that measure rather than the radius.

> ➤ **Review Video:** <u>Area and Circumference of a Circle</u>
> *Visit **mometrix.com/academy** and enter **Code**: 243015*

Concentric circles are circles that have the same center, but not the same length of radii. A bulls-eye target is an example of concentric circles.

An arc is a portion of a circle. Specifically, an arc is the set of points between and including two points on a circle. An arc does not contain any points inside the circle. When a segment is drawn from the endpoints of an arc to the center of the circle, a sector is formed.

A central angle is an angle whose vertex is the center of a circle and whose legs intercept an arc of the circle. Angle *XOY* in the diagram above is a central angle. A minor arc is an arc that has a measure less than 180°. The measure of a central angle is equal to the measure of the minor arc it intercepts. A major arc is an arc having a measure of at least 180°. The measure of the major arc can be found by subtracting the measure of the central angle from 360°.

A semicircle is an arc whose endpoints are the endpoints of the diameter of a circle. A semicircle is exactly half of a circle.

An inscribed angle is an angle whose vertex lies on a circle and whose legs contain chords of that circle. The portion of the circle intercepted by the legs of the angle is called the intercepted arc. The measure of the intercepted arc is exactly twice the measure of the inscribed angle. In the following diagram, angle *ABC* is an inscribed angle. $\overset{\frown}{AC} = 2(m\angle ABC)$

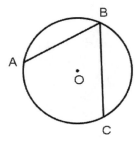

Any angle inscribed in a semicircle is a right angle. The intercepted arc is 180°, making the inscribed angle half that, or 90°. In the diagram below, angle *ABC* is inscribed in semicircle *ABC*, making angle *ABC* equal to 90°.

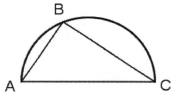

A chord is a line segment that has both endpoints on a circle. In the diagram below, $\overline{EB}$ is a chord. Secant: A line that passes through a circle and contains a chord of that circle. In the diagram below, $\overleftrightarrow{EB}$ is a secant and contains chord $\overline{EB}$.

A tangent is a line in the same plane as a circle that touches the circle in exactly one point. While a line segment can be tangent to a circle as part of a line that is tangent, it is improper to say a tangent can be simply a line segment that touches the circle in exactly one point. In the diagram below, $\overleftrightarrow{CD}$ is tangent to circle A. Notice that $\overline{FB}$ is not tangent to the circle. $\overline{FB}$ is a line segment that touches the circle in exactly one point, but if the segment were extended, it would touch the circle in a second point. The point at which a tangent touches a circle is called the point of tangency. In the diagram below, point B is the point of tangency.

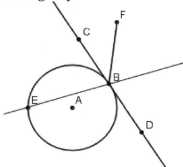

A secant is a line that intersects a circle in two points. Two secants may intersect inside the circle, on the circle, or outside the circle. When the two secants intersect on the circle, an inscribed angle is formed.

When two secants intersect inside a circle, the measure of each of two vertical angles is equal to half the sum of the two intercepted arcs. In the diagram below, $m\angle AEB = \frac{1}{2}(\widehat{AB} + \widehat{CD})$ and $m\angle BEC = \frac{1}{2}(\widehat{BC} + \widehat{AD})$.

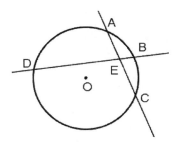

When two secants intersect outside a circle, the measure of the angle formed is equal to half the difference of the two arcs that lie between the two secants. In the diagram below, $m\angle E = \frac{1}{2}(\widehat{AB} - \widehat{CD})$.

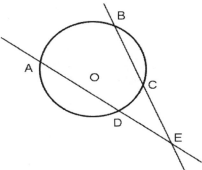

- 57 -

The arc length is the length of that portion of the circumference between two points on the circle. The formula for arc length is $s = \frac{\pi r \theta}{180°}$ where s is the arc length, r is the length of the radius, and θ is the angular measure of the arc in degrees, or $s = r\theta$, where θ is the angular measure of the arc in radians (2π radians = 360 degrees).

A sector is the portion of a circle formed by two radii and their intercepted arc. While the arc length is exclusively the points that are also on the circumference of the circle, the sector is the entire area bounded by the arc and the two radii.

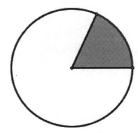

The area of a sector of a circle is found by the formula, $A = \frac{\theta r^2}{2}$, where A is the area, θ is the measure of the central angle in radians, and r is the radius. To find the area when the central angle is in degrees, use the formula, $A = \frac{\theta \pi r^2}{360}$, where θ is the measure of the central angle in degrees and r is the radius.

A circle is inscribed in a polygon if each of the sides of the polygon is tangent to the circle. A polygon is inscribed in a circle if each of the vertices of the polygon lies on the circle.

A circle is circumscribed about a polygon if each of the vertices of the polygon lies on the circle. A polygon is circumscribed about the circle if each of the sides of the polygon is tangent to the circle.

If one figure is inscribed in another, then the other figure is circumscribed about the first figure.

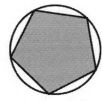

Circle circumscribed about a pentagon
Pentagon inscribed in a circle

Other Conic Sections

An ellipse is the set of all points in a plane, whose total distance from two fixed points called the foci (singular: focus) is constant, and whose center is the midpoint between the foci.

The standard equation of an ellipse that is taller than it is wide is $\frac{(y-k)^2}{a^2} + \frac{(x-h)^2}{b^2} = 1$, where a and b are coefficients. The center is the point (h, k) and the foci are the points $(h, k + c)$ and $(h, k - c)$, where $c^2 = a^2 - b^2$ and $a^2 > b^2$.

The major axis has length $2a$, and the minor axis has length $2b$.

Eccentricity (e) is a measure of how elongated an ellipse is, and is the ratio of the distance between the foci to the length of the major axis. Eccentricity will have a value between 0 and 1. The closer to 1 the eccentricity is, the closer the ellipse is to being a circle. The formula for eccentricity is $= \frac{c}{a}$.

Parabola: The set of all points in a plane that are equidistant from a fixed line, called the directrix, and a fixed point not on the line, called the focus.

Axis: The line perpendicular to the directrix that passes through the focus.

For parabolas that open up or down, the standard equation is $(x - h)^2 = 4c(y - k)$, where h, c, and k are coefficients. If c is positive, the parabola opens up. If c is negative, the parabola opens down. The vertex is the point (h, k). The directrix is the line having the equation $y = -c + k$, and the focus is the point $(h, c + k)$.

For parabolas that open left or right, the standard equation is $(y - k)^2 = 4c(x - h)$, where k, c, and h are coefficients. If c is positive, the parabola opens to the right. If c is negative, the parabola opens to the left. The vertex is the point (h, k). The directrix is the line having the equation $x = -c + h$, and the focus is the point $(c + h, k)$.

A hyperbola is the set of all points in a plane, whose distance from two fixed points, called foci, has a constant difference.

The standard equation of a horizontal hyperbola is $\frac{(x-h)^2}{a^2} - \frac{(y-k)^2}{b^2} = 1$, where a, b, h, and k are real numbers. The center is the point (h, k), the vertices are the points $(h + a, k)$ and $(h - a, k)$, and the foci are the points that every point on one of the parabolic curves is equidistant from and are found using the formulas $(h + c, k)$ and $(h - c, k)$, where $c^2 = a^2 + b^2$. The asymptotes are two lines the graph of the hyperbola approaches but never reaches, and are given by the equations $y = \left(\frac{b}{a}\right)(x - h) + k$ and $y = -\left(\frac{b}{a}\right)(x - h) + k$.

A vertical hyperbola is formed when a plane makes a vertical cut through two cones that are stacked vertex-to-vertex.

The standard equation of a vertical hyperbola is $\frac{(y-k)^2}{a^2} - \frac{(x-h)^2}{b^2} = 1$, where a, b, k, and h are real numbers. The center is the point (h, k), the vertices are the points $(h, k + a)$ and $(h, k - a)$, and the foci are the points that every point on one of the parabolic curves is equidistant from and are found using the formulas $(h, k + c)$ and $(h, k - c)$, where $c^2 = a^2 + b^2$. The asymptotes are two lines the graph of the hyperbola approaches but never reach, and are given by the equations $y = \left(\frac{a}{b}\right)(x - h) + k$ and $y = -\left(\frac{a}{b}\right)(x - h) + k$.

Solids

The surface area of a solid object is the area of all sides or exterior surfaces. For objects such as prisms and pyramids, a further distinction is made between base surface area (B) and lateral surface area (LA). For a prism, the total surface area (SA) is $SA = LA + 2B$. For a pyramid or cone, the total surface area is $SA = LA + B$.

> ➤ **Review Video:** <u>Finding Volume in Geometry</u>
> Visit *mometrix.com/academy* and enter *Code*: **754774**

The surface area of a sphere can be found by the formula $A = 4\pi r^2$, where r is the radius. The volume is given by the formula $V = \frac{4}{3}\pi r^3$, where r is the radius. Both quantities are generally given in terms of π.

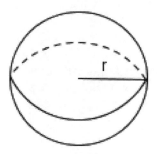

> ➤ **Review Video:** <u>Volume and Surface Area of a Sphere</u>
> Visit *mometrix.com/academy* and enter *Code*: **786928**

The volume of any prism is found by the formula $V = Bh$, where B is the area of the base, and h is the height (perpendicular distance between the bases). The surface area of any prism is the sum of the areas of both bases and all sides. It can be calculated as $SA = 2B + Ph$, where P is the perimeter of the base.

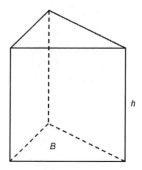

- 60 -

For a rectangular prism, the volume can be found by the formula $V = lwh$, where V is the volume, l is the length, w is the width, and h is the height. The surface area can be calculated as $SA = 2lw + 2hl + 2wh$ or $SA = 2(lw + hl + wh)$.

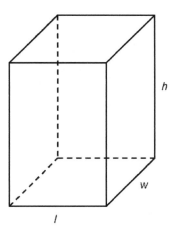

The volume of a cube can be found by the formula $V = s^3$, where s is the length of a side. The surface area of a cube is calculated as $SA = 6s^2$, where SA is the total surface area and s is the length of a side. These formulas are the same as the ones used for the volume and surface area of a rectangular prism, but simplified since all three quantities (length, width, and height) are the same.

> ➢ **Review Video: <u>Volume and Surface Area of a Cube</u>**
> *Visit **mometrix.com/academy** and enter **Code**: **664455**

The volume of a cylinder can be calculated by the formula $V = \pi r^2 h$, where r is the radius, and h is the height. The surface area of a cylinder can be found by the formula $SA = 2\pi r^2 + 2\pi rh$. The first term is the base area multiplied by two, and the second term is the perimeter of the base multiplied by the height.

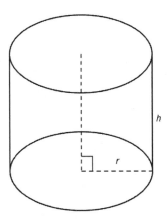

> ➢ **Review Video: <u>Volume and Surface Area of a Right Circular Cylinder</u>**
> *Visit **mometrix.com/academy** and enter **Code**: **226463**

The volume of a pyramid is found by the formula $V = \frac{1}{3}Bh$, where B is the area of the base, and h is the height (perpendicular distance from the vertex to the base). Notice this formula is the same as $\frac{1}{3}$ times the volume of a prism. Like a prism, the base of a pyramid can be any shape.

➢ **Review Video: <u>Volume and Surface Area of a Pyramid</u>**
Visit mometrix.com/academy and enter Code: **621932**

Finding the surface area of a pyramid is not as simple as the other shapes we've looked at thus far. If the pyramid is a right pyramid, meaning the base is a regular polygon and the vertex is directly over the center of that polygon, the surface area can be calculated as $SA = B + \frac{1}{2}Ph_s$, where P is the perimeter of the base, and h_s is the slant height (distance from the vertex to the midpoint of one side of the base). If the pyramid is irregular, the area of each triangle side must be calculated individually and then summed, along with the base.

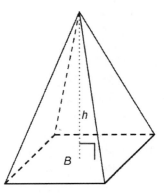

The volume of a cone is found by the formula $V = \frac{1}{3}\pi r^2 h$, where r is the radius, and h is the height. Notice this is the same as $\frac{1}{3}$ times the volume of a cylinder. The surface area can be calculated as $SA = \pi r^2 + \pi rs$, where s is the slant height. The slant height can be calculated using the Pythagorean Thereom to be $\sqrt{r^2 + h^2}$, so the surface area formula can also be written as $SA = \pi r^2 + \pi r\sqrt{r^2 + h^2}$.

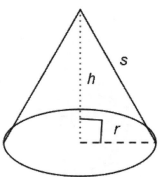

➢ **Review Video: <u>Volume and Surface Area of a Right Circular Cone</u>**
Visit mometrix.com/academy and enter Code: **573574**

Probability and Statistics

Probability

Probability is a branch of statistics that deals with the likelihood of something taking place. One classic example is a coin toss. There are only two possible results: heads or tails. The likelihood, or probability, that the coin will land as heads is 1 out of 2 (1/2, 0.5, 50%). Tails has the same probability. Another common example is a 6-sided die roll. There are six possible results from rolling a single die, each with an equal chance of happening, so the probability of any given number coming up is 1 out of 6.

Terms frequently used in probability:

Event – a situation that produces results of some sort (a coin toss)

Compound event – event that involves two or more independent events (rolling a pair of dice; taking the sum)

Outcome – a possible result in an experiment or event (heads, tails)

Desired outcome (or success) – an outcome that meets a particular set of criteria (a roll of 1 or 2 if we are looking for numbers less than 3)

Independent events – two or more events whose outcomes do not affect one another (two coins tossed at the same time)

Dependent events – two or more events whose outcomes affect one another (two cards drawn consecutively from the same deck)

Certain outcome – probability of outcome is 100% or 1

Impossible outcome – probability of outcome is 0% or 0

Mutually exclusive outcomes – two or more outcomes whose criteria cannot all be satisfied in a single event (a coin coming up heads and tails on the same toss)

Probability is the likelihood of a certain outcome occurring for a given event. The **theoretical probability** can usually be determined without actually performing the event. The likelihood of a outcome occurring, or the probability of an outcome occurring, is given by the formula

$$P(A) = \frac{\text{Number of acceptable outcomes}}{\text{Number of possible outcomes}}$$

where $P(A)$ is the probability of an outcome A occurring, and each outcome is just as likely to occur as any other outcome. If each outcome has the same probability of occurring as every other possible outcome, the outcomes are said to be equally likely to occur. The total number of acceptable outcomes must be less than or equal to the total number of possible outcomes. If the two are equal, then the outcome is certain to occur and the probability is 1. If the number of acceptable outcomes is zero, then the outcome is impossible and the probability is 0.

> ➤ **Review Video: <u>Theoretical and Experimental Probability</u>**
> *Visit **mometrix.com/academy** and enter **Code**: 466775*

Example:

There are 20 marbles in a bag and 5 are red. The theoretical probability of randomly selecting a red marble is 5 out of 20, (5/20 = 1/4, 0.25, or 25%).

When trying to calculate the probability of an event using the $\frac{desired\ outcomes}{total\ outcomes}$ formula, you may frequently find that there are too many outcomes to individually count them. Permutation and combination formulas offer a shortcut to counting outcomes. A permutation is an arrangement of a specific number of a set of objects in a specific order. The number of **permutations** of r items given a set of n items can be calculated as $_nP_r = \frac{n!}{(n-r)!}$. Combinations are similar to permutations, except there are no restrictions regarding the order of the elements. While ABC is considered a different permutation than BCA, ABC and BCA are considered the same combination. The number of **combinations** of r items given a set of n items can be calculated as $_nC_r = \frac{n!}{r!(n-r)!}$ or $_nC_r = \frac{_nP_r}{r!}$.

Example: Suppose you want to calculate how many different 5-card hands can be drawn from a deck of 52 cards. This is a combination since the order of the cards in a hand does not matter. There are 52 cards available, and 5 to be selected. Thus, the number of different hands is $_{52}C_5 = \frac{52!}{5! \times 47!} = 2,598,960$.

Sometimes it may be easier to calculate the possibility of something not happening, or the **complement of an event**. Represented by the symbol $\bar{A}$, the complement of A is the probability that event A does not happen. When you know the probability of event A occurring, you can use the formula $P(\bar{A}) = 1 - P(A)$, where $P(\bar{A})$ is the probability of event A not occurring, and $P(A)$ is the probability of event A occurring.

The **addition rule** for probability is used for finding the probability of a compound event. Use the formula $P(A\ or\ B) = P(A) + P(B) - P(A\ and\ B)$, where $P(A\ and\ B)$ is the probability of both events occurring to find the probability of a compound event. The probability of both events occurring at the same time must be subtracted to eliminate any overlap in the first two probabilities.

Conditional probability is the probability of an event occurring once another event has already occurred. Given event A and dependent event B, the probability of event B occurring when event A has already occurred is represented by the notation $P(A|B)$. To find the probability of event B occurring, take into account the fact that event A has already occurred and adjust the total number of possible outcomes. For example, suppose you have ten balls numbered 1–10 and you want ball number 7 to be pulled in two pulls. On the first pull, the probability of getting the 7 is $\frac{1}{10}$ because there is one ball with a 7 on it and 10 balls to choose from. Assuming the first pull did not yield a 7, the probability of pulling a 7 on the second pull is now $\frac{1}{9}$ because there are only 9 balls remaining for the second pull.

The **multiplication rule** can be used to find the probability of two independent events occurring using the formula $P(A\ and\ B) = P(A) \times P(B)$, where $P(A\ and\ B)$ is the probability of two independent events occurring, $P(A)$ is the probability of the first event occurring, and $P(B)$ is the probability of the second event occurring.

The multiplication rule can also be used to find the probability of two dependent events occurring using the formula $P(A\ and\ B) = P(A) \times P(B|A)$, where $P(A\ and\ B)$ is the probability of two dependent events occurring and $P(B|A)$ is the probability of the second event occurring after the first event has already occurred.

Before using the multiplication rule, you MUST first determine whether the two events are dependent or independent.

Use a combination of the multiplication rule and the rule of complements to find the probability that at least one outcome of the element will occur. This given by the general formula

$P(\text{at least one event occurring}) = 1 - P(\text{no outcomes occurring})$. For example, to find the probability that at least one even number will show when a pair of dice is rolled, find the probability that two odd numbers will be rolled (no even numbers) and subtract from one. You can always use a tree diagram or make a chart to list the possible outcomes when the sample space is small, such as in the dice-rolling example, but in most cases it will be much faster to use the multiplication and complement formulas.

Expected value is a method of determining expected outcome in a random situation. It is really a sum of the weighted probabilities of the possible outcomes. Multiply the probability of an event occurring by the weight assigned to that probability (such as the amount of money won or lost). A practical application of the expected value is to determine whether a game of chance is really fair. If the sum of the weighted probabilities is equal to zero, the game is generally considered fair because the player has a fair chance to at least to break even. If the expected value is less than zero, then players lose more than they win. For example, a lottery drawing might allow the player to choose any three-digit number, 000–999. The probability of choosing the winning number is 1:1000. If it costs $1 to play, and a winning number receives $500, the expected value is $\left(-\$1 \cdot \frac{999}{1,000}\right) + \left(\$500 \cdot \frac{1}{1,000}\right) = -0.499$ or $-\$0.50$. You can expect to lose on average 50 cents for every dollar you spend.

Most of the time, when we talk about probability, we mean theoretical probability. **Empirical probability**, or experimental probability or relative frequency, is the number of times an outcome occurs in a particular experiment or a certain number of observed events. While theoretical probability is based on what *should* happen, experimental probability is based on what *has* happened. Experimental probability is calculated in the same way as theoretical, except that actual outcomes are used instead of possible outcomes.

Theoretical and experimental probability do not always line up with one another. Theoretical probability says that out of 20 coin tosses, 10 should be heads. However, if we were actually to toss 20 coins, we might record just 5 heads. This doesn't mean that our theoretical probability is incorrect; it just means that this particular experiment had results that were different from what was predicted. A practical application of empirical probability is the insurance industry. There are no set functions that define life span, health, or safety. Insurance companies look at factors from hundreds of thousands of individuals to find patterns that they then use to set the formulas for insurance premiums.

Objective probability is based on mathematical formulas and documented evidence. Examples of objective probability include raffles or lottery drawings where there is a pre-determined number of possible outcomes and a predetermined number of outcomes that correspond to an event. Other cases of objective probability include probabilities of rolling dice, flipping coins, or drawing cards. Most gambling games are based on objective probability.

Subjective probability is based on personal or professional feelings and judgments. Often, there is a lot of guesswork following extensive research. Areas where subjective probability is applicable include sales trends and business expenses. Attractions set admission prices based on subjective probabilities of attendance based on varying admission rates in an effort to maximize their profit.

The total set of all possible results of a test or experiment is called a **sample space**, or sometimes a universal sample space. The sample space, represented by one of the variables S, Ω, or U (for universal sample space) has individual elements called outcomes. Other terms for outcome that may be used interchangeably include elementary outcome, simple event, or sample point. The number of outcomes in a given sample space could be infinite or finite, and some tests may yield multiple unique sample sets. For example, tests conducted by drawing playing cards from a standard deck would have one sample space of the card values, another sample space of the card suits, and a third sample space of suit-denomination combinations. For most tests, the sample spaces considered will be finite.

An event, represented by the variable E, is a portion of a sample space. It may be one outcome or a group of outcomes from the same sample space. If an event occurs, then the test or experiment will generate an outcome that satisfies the requirement of that event. For example, given a standard deck of 52 playing cards as the sample space, and defining the event as the collection of face cards, then the event will occur if the card drawn is a J, Q, or K. If any other card is drawn, the event is said to have not occurred.

For every sample space, each possible outcome has a specific likelihood, or probability, that it will occur. The probability measure, also called the distribution, is a function that assigns a real number probability, from zero to one, to each outcome. For a probability measure to be accurate, every outcome must have a real number probability measure that is greater than or equal to zero and less than or equal to one. Also, the probability measure of the sample space must equal one, and the probability measure of the union of multiple outcomes must equal the sum of the individual probability measures.

Probabilities of events are expressed as real numbers from zero to one. They give a numerical value to the chance that a particular event will occur. The probability of an event occurring is the sum of the probabilities of the individual elements of that event. For example, in a standard deck of 52 playing cards as the sample space and the collection of face cards as the event, the probability of drawing a specific face card is $\frac{1}{52} = 0.019$, but the probability of drawing any one of the twelve face cards is $12(0.019) = 0.228$. Note that rounding of numbers can generate different results. If you multiplied 12 by the fraction $\frac{1}{52}$ before converting to a decimal, you would get the answer $\frac{12}{52} = 0.231$.

For a simple sample space, possible outcomes may be determined by using a **tree diagram** or an organized chart. In either case, you can easily draw or list out the possible outcomes. For example, to determine all the possible ways three objects can be ordered, you can draw a tree diagram:

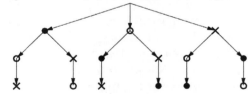

You can also make a chart to list all the possibilities:

First object	Second object	Third object
●	X	O
●	O	X
O	●	X
O	X	●
X	●	O
X	O	●

Either way, you can easily see there are six possible ways the three objects can be ordered.

If two events have no outcomes in common, they are said to be mutually exclusive. For example, in a standard deck of 52 playing cards, the event of all card suits is mutually exclusive to the event of all card values. If two events have no bearing on each other so that one event occurring has no influence on the probability of another event occurring, the two events are said to be independent. For example, rolling a standard six-sided die multiple times does not change that probability that a particular number will be rolled from one roll to the next. If the outcome of one event does affect the probability of the second event, the two events are said to be dependent. For example, if cards are drawn from a deck, the probability of drawing an ace after an ace has been drawn is different than the probability of drawing an ace if no ace (or no other card, for that matter) has been drawn.

In probability, the odds in favor of an event are the number of times the event will occur compared to the number of times the event will not occur. To calculate the odds in favor of an event, use the formula $\frac{P(A)}{1-P(A)}$, where $P(A)$ is the probability that the event will occur. Many times, odds in favor is given as a ratio in the form $\frac{a}{b}$ or $a{:}b$, where a is the probability of the event occurring and b is the complement of the event, the probability of the event not occurring. If the odds in favor are given as 2:5, that means that you can expect the event to occur two times for every 5 times that it does not occur. In other words, the probability that the event will occur is $\frac{2}{2+5} = \frac{2}{7}$.

In probability, the odds against an event are the number of times the event will not occur compared to the number of times the event will occur. To calculate the odds against an event, use the formula $\frac{1-P(A)}{P(A)}$, where $P(A)$ is the probability that the event will occur. Many times, odds against is given as a ratio in the form $\frac{b}{a}$ or $b{:}a$, where b is the probability the event will not occur (the complement of the event) and a is the probability the event will occur. If the odds against an event are given as 3:1, that means that you can expect the event to not occur 3 times for every one time it does occur. In other words, 3 out of every 4 trials will fail.

Statistics

In statistics, the *Population* is the entire collection of people, plants, etc., that data can be collected from. For example, a study to determine how well students in the area schools perform on a standardized test would have a population of all the students enrolled in those schools, although a study may include just a small sample of students from each school. A *Parameter* is a numerical value that gives information about the population, such as the mean, median, mode, or standard deviation. Remember that the symbol for the mean of a population is μ and the symbol for the standard deviation of a population is σ.

A *Sample* is a portion of the entire population. Where as a parameter helped describe the population, a *Statistic* is a numerical value that gives information about the sample, such as mean, median, mode, or standard deviation. Keep in mind that the symbols for mean and standard deviation are different when they are referring to a sample rather than the entire population. For a sample, the symbol for mean is $\bar{x}$ and the symbol for standard deviation is s. The mean and standard deviation of a sample may or may not be identical to that of the entire population due to a sample only being a subset of the population. However, if the sample is random and large enough, statistically significant values can be attained. Samples are generally used when the population is too large to justify including every element or when acquiring data for the entire population is impossible.

Inferential Statistics is the branch of statistics that uses samples to make predictions about an entire population. This type of statistics is often seen in political polls, where a sample of the population is questioned about a particular topic or politician to gain an understanding about the attitudes of the entire population of the country. Often, exit polls are conducted on election days using this method. Inferential statistics can have a large margin of error if you do not have a valid sample.

Statistical values calculated from various samples of the same size make up the sampling distribution. For example, if several samples of identical size are randomly selected from a large population and then the mean of each sample is calculated, the distribution of values of the means would be a *Sampling Distribution*.

The *Sampling Distribution of the Mean* is the distribution of the sample mean, $\bar{x}$, derived from random samples of a given size. It has three important characteristics. First, the mean of the sampling distribution of the mean is equal to the mean of the population that was sampled. Second, assuming the standard deviation is non-zero, the standard deviation of the sampling distribution of the mean equals the standard deviation of the sampled population divided by the square root of the sample size. This is sometimes called the standard error. Finally, as the sample size gets larger, the sampling distribution of the mean gets closer to a normal distribution via the Central Limit Theorem.

A *Survey Study* is a method of gathering information from a small group in an attempt to gain enough information to make accurate general assumptions about the population. Once a survey study is completed, the results are then put into a summary report.
Survey studies are generally in the format of surveys, interviews, or questionnaires as part of an effort to find opinions of a particular group or to find facts about a group.
It is important to note that the findings from a survey study are only as accurate as the sample chosen from the population.

Correlational Studies seek to determine how much one variable is affected by changes in a second variable. For example, correlational studies may look for a relationship between the amount of time a student spends studying for a test and the grade that student earned on the test or between student scores on college admissions tests and student grades in college.
It is important to note that correlational studies cannot show a cause and effect, but rather can show only that two variables are or are not potentially correlated.

Experimental Studies take correlational studies one step farther, in that they attempt to prove or disprove a cause-and-effect relationship. These studies are performed by conducting a series of experiments to test the hypothesis. For a study to be scientifically accurate, it must have both an experimental group that receives the specified treatment and a control group that does not get the treatment. This is the type of study pharmaceutical companies do as part of drug trials for new medications. Experimental studies are only valid when proper scientific method has been followed. In other words, the experiment must be well-planned and executed without bias in the testing process, all subjects must be selected at random, and the process of determining which subject is in which of the two groups must also be completely random.

Observational Studies are the opposite of experimental studies. In observational studies, the tester cannot change or in any way control all of the variables in the test. For example, a study to determine which gender does better in math classes in school is strictly observational. You cannot change a person's gender, and you cannot change the subject being studied. The big downfall of observational studies is that you have no way of proving a cause-and-effect relationship because you cannot control outside influences. Events outside of school can influence a student's performance in school, and observational studies cannot take that into consideration.

For most studies, a *Random Sample* is necessary to produce valid results. Random samples should not have any particular influence to cause sampled subjects to behave one way or another. The goal is for the random sample to be a *Representative Sample*, or a sample whose characteristics give an accurate picture of the characteristics of the entire population. To accomplish this, you must make sure you have a proper *Sample Size*, or an appropriate number of elements in the sample.

In statistical studies, biases must be avoided. *Bias* is an error that causes the study to favor one set of results over another. For example, if a survey to determine how the country views the president's job performance only speaks to registered voters in the president's party, the results will be skewed because a disproportionately large number of responders would tend to show approval, while a disproportionately large number of people in the opposite party would tend to express disapproval.

Extraneous Variables are, as the name implies, outside influences that can affect the outcome of a study. They are not always avoidable, but could trigger bias in the result.

Data Analysis

The *Measure of Central Tendency* is a statistical value that gives a general tendency for the center of a group of data. There are several different ways of describing the measure of central tendency. Each one has a unique way it is calculated, and each one gives a slightly different perspective on the data set. Whenever you give a measure of central tendency, always make sure the units are the same. If the data has different units, such as hours, minutes, and seconds, convert all the data to the same unit, and use the same unit in the measure of central tendency. If no units are given in the data, do not give units for the measure of central tendency.

The statistical *Mean* of a group of data is the same as the arithmetic average of that group. To find the mean of a set of data, first convert each value to the same units, if necessary. Then find the sum of all the values, and count the total number of data values, making sure you take into consideration each individual value. If a value appears more than once, count it more than once. Divide the sum of the values by the total number of values and apply the units, if any. Note that the mean does not have to be one of the data values in the set, and may not divide evenly.

$$\text{mean} = \frac{\text{sum of the data values}}{\text{quantity of data values}}$$

While the mean is relatively easy to calculate and averages are understood by most people, the mean can be very misleading if used as the sole measure of central tendency. If the data set has outliers (data values that are unusually high or unusually low compared to the rest of the data values), the mean can be very distorted, especially if the data set has a small number of values. If unusually high values are countered with unusually low values, the mean is not affected as much. For example, if five of twenty students in a class get a 100 on a test, but the other 15 students have an average of 60 on the same test, the class average would appear as 70. Whenever the mean is skewed by outliers, it is always a good idea to include the median as an alternate measure of central tendency.

The statistical *Median* is the value in the middle of the set of data. To find the median, list all data values in order from smallest to largest or from largest to smallest. Any value that is repeated in the set must be listed the number of times it appears. If there are an odd number of data values, the median is the value in the middle of the list. If there is an even number of data values, the median is the arithmetic mean of the two middle values.

The statistical *Mode* is the data value that occurs the most number of times in the data set. It is possible to have exactly one mode, more than one mode, or no mode. To find the mode of a set of data, arrange the data like you do to find the median (all values in order, listing all multiples of data values). Count the number of times each value appears in the data set. If all values appear an equal number of times, there is no mode. If one value appears more than any other value, that value is the mode. If two or more values appear the same number of times, but there are other values that appear fewer times and no values that appear more times, all of those values are the modes.

The big disadvantage of using the median as a measure of central tendency is that is relies solely on a value's relative size as compared to the other values in the set. When the individual values in a set of data are evenly dispersed, the median can be an accurate tool. However, if there is a group of rather large values or a group of rather small values that are not offset by a different group of values, the information that can be inferred from the median may not be accurate because the distribution of values is skewed.

The main disadvantage of the mode is that the values of the other data in the set have no bearing on the mode. The mode may be the largest value, the smallest value, or a value anywhere in between in the set. The mode only tells which value or values, if any, occurred the most number of times. It does not give any suggestions about the remaining values in the set.

The *Measure of Dispersion* is a single value that helps to "interpret" the measure of central tendency by providing more information about how the data values in the set are distributed about the measure of central tendency. The measure of dispersion helps to eliminate or reduce the disadvantages of using the mean, median, or mode as a single measure of central tendency, and give a more accurate picture of the data set as a whole. To have a measure of dispersion, you must know or calculate the range, standard deviation, or variance of the data set.

The *Range* of a set of data is the difference between the greatest and lowest values of the data in the set. To calculate the range, you must first make sure the units for all data values are the same, and then identify the greatest and lowest values. Use the formula range = highest value – lowest value. If there are multiple data values that are equal for the highest or lowest, just use one of the values in the formula. Write the answer with the same units as the data values you used to do the calculations.

Standard Deviation is a measure of dispersion that compares all the data values in the set to the mean of the set to give a more accurate picture. To find the standard deviation of a population, use the formula

$$\sigma = \sqrt{\frac{\sum_{i=1}^{n}(x_i - \bar{x})^2}{n}}$$

where σ is the standard deviation of a population, x represents the individual values in the data set, $\bar{x}$ is the mean of the data values in the set, and n is the number of data values in the set. The higher the value of the standard deviation is, the greater the variance of the data values from the mean. The units associated with the standard deviation are the same as the units of the data values.

The *Variance* of a population, or just variance, is the square of the standard deviation of that population. While the mean of a set of data gives the average of the set and gives information about where a specific data value lies in relation to the average, the variance of the population gives information about the degree to which the data values are spread out and tell you how close an individual value is to the average compared to the other values. The units associated with variance are the same as the units of the data values squared.

Percentiles and Quartiles are other methods of describing data within a set. *Percentiles* tell what percentage of the data in the set fall below a specific point. For example, achievement test scores are often given in percentiles. A score at the 80th percentile is one which is equal to or higher than 80 percent of the scores in the set. In other words, 80 percent of the scores were lower than that score.

Quartiles are percentile groups that make up quarter sections of the data set. The first quartile is the 25th percentile. The second quartile is the 50th percentile; this is also the median of the data set. The third quartile is the 75th percentile.

Skewness is a way to describe the symmetry or asymmetry of the distribution of values in a data set. If the distribution of values is symmetrical, there is no skew. In general the closer the mean of a data set is to the median of the data set, the less skew there is. Generally, if the mean is to the right of the median, the data set is *Positively Skewed*, or right-skewed, and if the mean is to the left of the median, the data set is *negatively skewed*, or left-skewed. However, this rule of thumb is not infallible. When the data values are graphed on a curve, a set with no skew will be a perfect bell curve. To estimate skew, use the formula

$$\text{skew} = \frac{\sqrt{n(n-1)}}{n-2}\left(\frac{\frac{1}{n}\sum_{i=1}^{n}(x_i - \bar{x})^3}{\left(\frac{1}{n}\sum_{i=1}^{n}(x_i - \bar{x})^2\right)^{\frac{3}{2}}}\right)$$

where n is the number of values is the set, x_i is the ith value in the set, and $\bar{x}$ is the mean of the set.

In statistics, *Simple Regression* is using an equation to represent a relation between an independent and dependent variables. The independent variable is also referred to as the explanatory variable or the predictor, and is generally represented by the variable x in the equation. The dependent variable, usually represented by the variable y, is also referred to as the response variable. The equation may be any type of function – linear, quadratic, exponential, etc. The best way to handle this task is to use the regression feature of your graphing calculator. This will easily give you the curve of best fit and provide you with the coefficients and other information you need to derive an equation.

In a scatter plot, the *Line of Best Fit* is the line that best shows the trends of the data. The line of best fit is given by the equation $\hat{y} = ax + b$, where a and b are the regression coefficients. The regression coefficient a is also the slope of the line of best fit, and b is also the y-coordinate of the point at which the line of best fit crosses the x-axis. Not every point on the scatter plot will be on the line of best fit. The differences between the y-values of the points in the scatter plot and the corresponding y-values according to the equation of the line of best fit are the residuals. The line of best fit is also called the least-squares regression line because it is also the line that has the lowest sum of the squares of the residuals.

The *Correlation Coefficient* is the numerical value that indicates how strong the relationship is between the two variables of a linear regression equation. A correlation coefficient of –1 is a perfect negative correlation. A correlation coefficient of +1 is a perfect positive correlation. Correlation coefficients close to –1 or +1 are very strong correlations. A correlation coefficient equal to zero indicates there is no correlation between the two variables. This test is a good indicator of whether or not the equation for the line of best fit is accurate. The formula for the correlation coefficient is

$$r = \frac{\sum_{i=1}^{n}(x_i - \bar{x})(y_i - \bar{y})}{\sqrt{\sum_{i=1}^{n}(x_i - \bar{x})^2}\sqrt{\sum_{i=1}^{n}(y_i - \bar{y})^2}}$$

where r is the correlation coefficient, n is the number of data values in the set, (x_i, y_i) is a point in the set, and $\bar{x}$ and $\bar{y}$ are the means.

A *Z-score* is an indication of how many standard deviations a given value falls from the mean. To calculate a z-score, use the formula $= \frac{x-\mu}{\sigma}$, where x is the data value, μ is the mean of the data set, and σ is the standard deviation of the population. If the z-score is positive, the data value lies above the mean. If the z-score is negative, the data value falls below the mean. These scores are useful in interpreting data such as standardized test scores, where every piece of data in the set has been counted, rather than just a small random sample. In cases where standard deviations are calculated from a random sample of the set, the z-scores will not be as accurate.

According to the *Central Limit Theorem*, regardless of what the original distribution of a sample is, the distribution of the means tends to get closer and closer to a normal distribution as the sample size gets larger and larger (this is necessary because the sample is becoming more all-encompassing of the elements of the population). As the sample size gets larger, the distribution of the sample mean will approach a normal distribution with a mean of the population mean and a variance of the population variance divided by the sample size.

Displaying Information

Charts and *Tables* are ways of organizing information into separate rows and columns that are labeled to identify and explain the data contained in them. Some charts and tables are organized horizontally, with row lengths giving the details about the labeled information. Other charts and tables are organized vertically, with column heights giving the details about the labeled information.

Frequency Tables show how frequently each unique value appears in the set. A *Relative Frequency Table* is one that shows the proportions of each unique value compared to the entire set. Relative frequencies are given as percents; however, the total percent for a relative frequency table will not necessarily equal 100 percent due to rounding. An example of a frequency table with relative frequencies is below.

Favorite Color	Frequency	Relative Frequency
Blue	4	13%
Red	7	22%
Purple	3	9%
Green	6	19%
Cyan	12	38%

A *Pictograph* is a graph, generally in the horizontal orientation, that uses pictures or symbols to represent the data. Each pictograph must have a key that defines the picture or symbol and gives the quantity each picture or symbol represents. Pictures or symbols on a pictograph are not always shown as whole elements. In this case, the fraction of the picture or symbol shown represents the same fraction of the quantity a whole picture or symbol stands for. For example, a row with $3\frac{1}{2}$ ears of corn, where each ear of corn represents 100 stalks of corn in a field, would equal $3\frac{1}{2} \cdot 100 = 350$ stalks of corn in the field.

Circle Graphs, also known as *Pie Charts*, provide a visual depiction of the relationship of each type of data compared to the whole set of data. The circle graph is divided into sections by drawing radii to create central angles whose percentage of the circle is equal to the individual data's percentage of the whole set. Each 1% of data is equal to 3.6° in the circle graph. Therefore, data represented by a 90° section of the circle graph makes up 25% of the whole. When complete, a circle graph often looks like a pie cut into uneven wedges. The pie chart below shows the data from the frequency table referenced earlier where people were asked their favorite color.

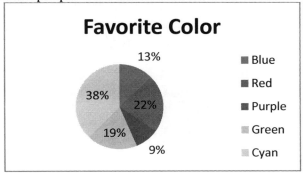

> ➤ **Review Video: Pie Chart**
> *Visit* **mometrix.com/academy** *and enter* **Code: 895285**

Line Graphs have one or more lines of varying styles (solid or broken) to show the different values for a set of data. The individual data are represented as ordered pairs, much like on a Cartesian plane. In this case, the *x*- and *y*- axes are defined in terms of their units, such as dollars or time. The individual plotted points are joined by line segments to show whether the value of the data is increasing (line sloping upward), decreasing (line sloping downward) or staying the same (horizontal line). Multiple sets of data can be graphed on the same line graph to give an easy visual comparison. An example of this would be graphing achievement test scores for different groups of students over the same time period to see which group had the greatest increase or decrease in performance from year-to-year (as shown below).

> **Review Video: Line Graphs**
> Visit *mometrix.com/academy* and enter *Code*: **480147**

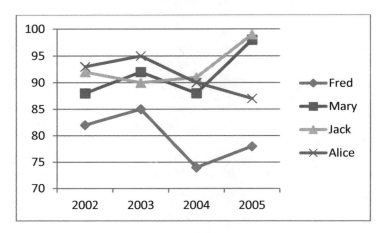

A *Line Plot*, also known as a *Dot Plot*, has plotted points that are NOT connected by line segments. In this graph, the horizontal axis lists the different possible values for the data, and the vertical axis lists the number of times the individual value occurs. A single dot is graphed for each value to show the number of times it occurs. This graph is more closely related to a bar graph than a line graph. Do not connect the dots in a line plot or it will misrepresent the data.

> **Review Video: Line Plot**
> Visit *mometrix.com/academy* and enter *Code*: **754610**

A *Stem and Leaf Plot* is useful for depicting groups of data that fall into a range of values. Each piece of data is separated into two parts: the first, or left, part is called the stem; the second, or right, part is called the leaf. Each stem is listed in a column from smallest to largest. Each leaf that has the common stem is listed in that stem's row from smallest to largest. For example, in a set of two-digit numbers, the digit in the tens place is the stem, and the digit in the ones place is the leaf. With a stem and leaf plot, you can easily see which subset of numbers (10s, 20s, 30s, etc.) is the largest. This information is also readily available by looking at a histogram, but a stem and leaf plot also allows you to look closer and see exactly which values fall in that range. Using all of the test scores from above, we can assemble a stem and leaf plot like the one below.

Test Scores								
7	4	8						
8	2	5	7	8	8			
9	0	0	1	2	2	3	5	8	9

A *Bar Graph* is one of the few graphs that can be drawn correctly in two different configurations – both horizontally and vertically. A bar graph is similar to a line plot in the way the data is organized on the graph. Both axes must have their categories defined for the graph to be useful. Rather than placing a single dot to mark the point of the data's value, a bar, or thick line, is drawn from zero to the exact value of the data, whether it is a number, percentage, or other numerical value. Longer bar lengths correspond to greater data values. To read a bar graph, read the labels for the axes to find the units being reported. Then look where the bars end in relation to the scale given on the corresponding axis and determine the associated value.

The bar chart below represents the responses from our favorite color survey.

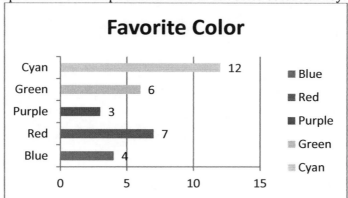

> **Review Video: Bar Graph**
> Visit *mometrix.com/academy* and enter *Code*: **226729**

At first glance, a *Histogram* looks like a vertical bar graph. The difference is that a bar graph has a separate bar for each piece of data and a histogram has one continuous bar for each *Range* of data. For example, a histogram may have one bar for the range 0–9, one bar for 10–19, etc. While a bar graph has numerical values on one axis, a histogram has numerical values on both axes. Each range is of equal size, and they are ordered left to right from lowest to highest. The height of each column on a histogram represents the number of data values within that range. Like a stem and leaf plot, a histogram makes it easy to glance at the graph and quickly determine which range has the greatest quantity of values. A simple example of a histogram is below.

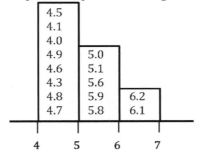

> **Review Video: Histogram**
> Visit *mometrix.com/academy* and enter *Code*: **735897**

Bivariate Data is simply data from two different variables. (The prefix *bi-* means *two*.) In a *Scatter Plot*, each value in the set of data is plotted on a grid similar to a Cartesian plane, where each axis represents one of the two variables. By looking at the pattern formed by the points on the grid, you can often determine whether or not there is a relationship between the two variables, and what that relationship is, if it exists. The variables may be directly proportionate, inversely proportionate, or show no proportion at all. It may also be possible to determine if the data is linear, and if so, to find an equation to relate the two variables. The following scatter plot shows the relationship between preference for brand "A" and the age of the consumers surveyed.

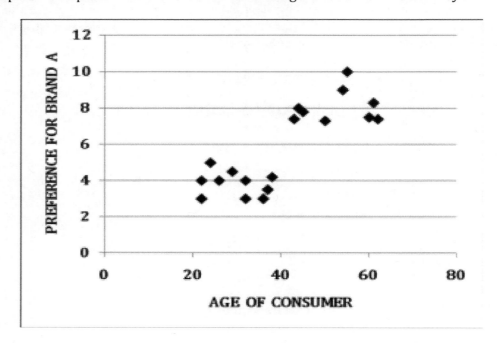

Scatter Plots are also useful in determining the type of function represented by the data and finding the simple regression. Linear scatter plots may be positive or negative. Nonlinear scatter plots are generally exponential or quadratic. Below are some common types of scatter plots:

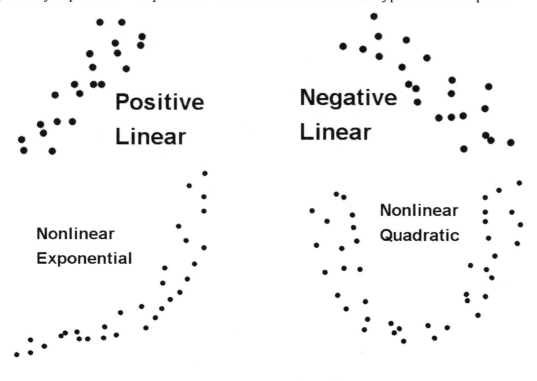

➤ **Review Video: Scatter Plot**
Visit *mometrix.com/academy* and enter **Code: 596526**

The *5-Number Summary* of a set of data gives a very informative picture of the set. The five numbers in the summary include the minimum value, maximum value, and the three quartiles. This information gives the reader the range and median of the set, as well as an indication of how the data is spread about the median.

A *Box-and-Whiskers Plot* is a graphical representation of the 5-number summary. To draw a box-and-whiskers plot, plot the points of the 5-number summary on a number line. Draw a box whose ends are through the points for the first and third quartiles. Draw a vertical line in the box through the median to divide the box in half. Draw a line segment from the first quartile point to the minimum value, and from the third quartile point to the maximum value.

The *68–95–99.7 Rule* describes how a normal distribution of data should appear when compared to the mean. This is also a description of a normal bell curve. According to this rule, 68 percent of the data values in a normally distributed set should fall within one standard deviation of the mean (34 percent above and 34 percent below the mean), 95 percent of the data values should fall within two standard deviations of the mean (47.5 percent above and 47.5 percent below the mean), and 99.7 percent of the data values should fall within three standard deviations of the mean, again, equally distributed on either side of the mean. This means that only 0.3 percent of all data values should fall more than three standard deviations from the mean. On the graph below, the normal curve is centered on the y-axis. The x-axis labels are how many standard deviations away from the center you are.

Therefore, it is easy to see how the 68-95-99.7 rule can apply.

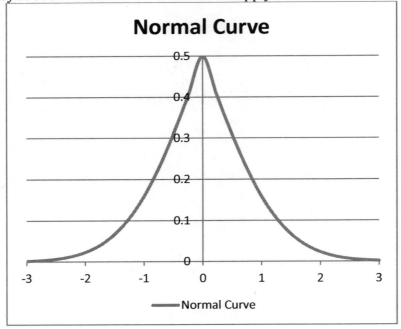

The five general shapes of frequency curves are *Symmetrical*, *U-shaped*, *Skewed*, *J-shaped*, and *Multimodal*. Symmetrical curves are also known as bell curves or normal curves. Values equidistant from the median have equal frequencies. U-shaped curves have two maxima – one at each end. Skewed curves have the maximum point off-center. Curves that are negative skewed, or left skewed, have the maximum on the right side of the graph so there is longer tail and lower slope on the left side. The opposite is true for curves that are positive skewed, or right skewed. J-shaped curves have a maximum at one end and a minimum at the other end. Multimodal curves have multiple maxima. For example, if the curve has exactly two maxima, it is called a bimodal curve.

Discrete Mathematics

Factorials

The **factorial** is a function that can be performed on any **non-negative integer**. It is represented by the ! sign written after the integer on which it is being performed. The factorial of an integer is the product of all positive integers less than or equal to the number. For example, 4! (read "4 factorial") is calculated as $4 \times 3 \times 2 \times 1 = 24$.

Since 0 is not itself a positive integer, nor does it have any positive integers less than it, 0! cannot be calculated using this method. Instead, 0! is defined by convention to equal 1. This makes sense if you consider the pattern of descending factorials:

$$5! = 120$$
$$4! = \frac{5!}{5} = 24$$
$$3! = \frac{4!}{4} = 6$$
$$2! = \frac{3!}{3} = 2$$
$$1! = \frac{2!}{2} = 1$$
$$0! = \frac{1!}{1} = 1$$

Permutations and Combinations

For any given set of data, the individual elements in the set may be arranged in different groups containing different numbers of elements arranged in different orders. For example, given the set of integers from one to three, inclusive, the elements of the set are 1, 2, and 3: written as {1, 2, 3}. They may be arranged as follows: 1, 2, 3, 12, 21, 13, 31, 23, 32, 123, 132, 213, 231, 312, and 321. These ordered sequences of elements from the given set of data are called **permutations**. It is important to note that in permutations, the order of the elements in the sequence is important. The sequence 123 is not the same as the sequence 213. Also, no element in the given set may be used more times as an element in a permutation than it appears as an element in the original set. For example, 223 is not a permutation in the above example because the number 2 only appears one time in the given set.

To find the number of permutations of r items from a set of n items, use the formula $_nP_r = \frac{n!}{(n-r)!}$. When using this formula, each element of r must be unique. Also, this assumes that different arrangements of the same set of elements yields different outcomes. For example, 123 is not the same as 321; order is important.

A special case arises while finding the number of possible permutations of n items from a set of n items. Because $n = r$, the equation for the number of permutations becomes simply $P = n!$.

If a set contains one or more groups of **indistinguishable or interchangeable elements** (e.g., the set {1, 2, 3, 3}, which has a group of two indistinguishable 3's), there is a different formula for

- 79 -

finding distinct permutations of all n elements. Use the formula $P = \dfrac{n!}{m_1!m_2!...m_k!}$, where P is the number of permutations, n is the total number of elements in the set, and m_1 through m_k are the number of identical elements in each group (e.g., for the set {1, 1, 2, 2, 2, 3, 3}, $m_1 = 2$, $m_2 = 3$, and $m_3 = 2$). It is important to note that each repeated number is counted as its own element for the purpose of defining n (e.g., for the set {1, 1, 2, 2, 2, 3, 3}, $n = 7$, not 3).

To find the number of possible permutations of **any number of elements** in a set of unique elements, you must apply the permutation formulas multiple times. For example, to find the total number of possible permutations of the set {1, 2, 3} first apply the permutation formula for situations where $n = r$ as follows: $P = n! = 3! = 6$. This gives the number of permutations of the three elements when all three elements are used. To find the number of permutations when only two of the three elements are used, use the formula $_nP_r = \dfrac{n!}{(n-r)!}$, where n is 3 and r is 2.

$$_nP_r = \frac{n!}{(n-r)!} \Rightarrow {}_3P_2 = \frac{3!}{(3-2)!} = \frac{6}{1} = 6$$

To find the number of permutations when one element is used, use the formula $_nP_r = \dfrac{n!}{(n-r)!}$, where n is 3 and r is 1.

$$_nP_r = \frac{n!}{(n-r)!} \Rightarrow {}_3P_1 = \frac{3!}{(3-1)!} = \frac{3!}{2!} = \frac{6}{2} = 3$$

Find the sum of the three formulas: $6 + 6 + 3 = 15$ total possible permutations.

Alternatively, the general formula for total possible permutations can be written as follows:

$$P_T = \sum_{i=1}^{n} \frac{n!}{(i-1)!}$$

Combinations are essentially defined as permutations where the order in which the elements appear does not matter. Going back to the earlier example of the set {1, 2, 3}, the possible combinations that can be made from that set are 1, 2, 3, 12, 13, 23, and 123.

In a set containing n elements, the number of combinations of r items from the set can be found using the formula $_nC_r = \dfrac{n!}{r!(n-r)!}$. Notice the similarity to the formula for permutations. In effect, you are dividing the number of permutations by $r!$ to get the number of combinations, and the formula may be written $_nC_r = \dfrac{{}_nP_r}{r!}$. When finding the number of combinations, it is important to remember that the elements in the set must be unique (i.e., there must not be any duplicate items), and that no item may be used more than once in any given sequence.

Sequences

A sequence is a set of numbers that continues on in a define pattern. The function that defines a sequence has a domain composed of the set of positive integers. Each member of the sequence is an element, or individual term. Each element is identified by the notation a_n, where a is the term of the sequence, and n is the integer identifying which term in the sequence a is. There are two different ways to represent a sequence that contains the element a_n. The first is the simple notation $\{a_n\}$. The expanded notation of a sequence is $a_1, a_2, a_3, \ldots a_n, \ldots$. Notice that the expanded form does not end with the nth term. There is no indication that the nth term is the last term in the sequence, only that the nth term is an element of the sequence.

Some sequences will have a limit, or a value the sequence approaches or sometimes even reaches but never passes. A sequence that has a limit is known as a convergent sequence because all the values of the sequence seemingly converge at that point. Sequences that do not converge at a particular limit are divergent sequences. The easiest way to determine whether a sequence converges or diverges is to find the limit of the sequence. If the limit is a real number, the sequence is a convergent sequence. If the limit is infinity, the sequence is a divergent sequence.

Remember the following rules for finding limits:
$\lim_{n\to\infty} k = k$ for all real numbers k
$\lim_{n\to\infty} \frac{1}{n} = 0$
$\lim_{n\to\infty} n = \infty$
$\lim_{n\to\infty} \frac{k}{n^p} = 0$ for all real numbers k and positive rational numbers p.

The limit of the sums of two sequences is equal to the sum of the limits of the two sequences:
$\lim_{n\to\infty}(a_n + b_n) = \lim_{n\to\infty} a_n + \lim_{n\to\infty} b_n$.

The limit of the difference between two sequences is equal to the difference between the limits of the two sequences:
$\lim_{n\to\infty}(a_n - b_n) = \lim_{n\to\infty} a_n - \lim_{n\to\infty} b_n$.

The limit of the product of two sequences is equal to the product of the limits of the two sequences:
$\lim_{n\to\infty}(a_n \cdot b_n) = \lim_{n\to\infty} a_n \cdot \lim_{n\to\infty} b_n$.

The limit of the quotient of two sequences is equal to the quotient of the limits of the two sequences, with some exceptions: $\lim_{n\to\infty}\left(\frac{a_n}{b_n}\right) = \frac{\lim_{n\to\infty} a_n}{\lim_{n\to\infty} b_n}$. In the quotient formula, it is important to consider that $b_n \neq 0$ and $\lim_{n\to\infty} b_n \neq 0$.

The limit of a sequence multiplied by a scalar is equal to the scalar multiplied by the limit of the sequence: $\lim_{n\to\infty} ka_n = k \lim_{n\to\infty} a_n$, where k is any real number.

A **monotonic sequence** is a sequence that is either nonincreasing or nondecreasing. The term *nonincreasing* is used to describe a sequence whose terms either get progressively smaller in value or remain the same. The term *nondecreasing* is used to describe a sequence whose terms either get progressively larger in value or remain the same. A nonincreasing sequence is bounded above. This means that all elements of the sequence must be less than a given real number. A nondecreasing sequence is bounded below. This means that all elements of the sequence must be greater than a given real number.

When one element of a sequence is defined in terms of a previous element or elements of the sequence, the sequence is a **recursive sequence**. For example, given the recursive definition $a_1 = 0$; $a_2 = 1$; $a_n = a_{n-1} + a_{n-2}$ for all $n \geq 2$, you get the sequence 0, 1, 1, 2, 3, 5, 8, … . This particular sequence is known as the Fibonacci sequence, and is defined as the numbers zero and one, and a continuing sequence of numbers, with each number in the sequence equal to the sum of

- 81 -

the two previous numbers. It is important to note that the Fibonacci sequence can also be defined as the first two terms being equal to one, with the remaining terms equal to the sum of the previous two terms. Both definitions are considered correct in mathematics. Make sure you know which definition you are working with when dealing with Fibonacci numbers.

Sometimes one term of a sequence with a recursive definition can be found without knowing the previous terms of the sequence. This case is known as a closed-form expression for a recursive definition. In this case, an alternate formula will apply to the sequence to generate the same sequence of numbers. However, not all sequences based on recursive definitions will have a closed-form expression. Some sequences will require the use of the recursive definition. For example, the Fibonacci sequence has a closed-form expression given by the formula $a_n = \frac{\phi^n - \left(\frac{-1}{\phi}\right)^n}{\sqrt{5}}$, where φ is the golden ratio, which is equal to $\frac{1+\sqrt{5}}{2}$. In this case, $a_0 = 0$ and $a_1 = 1$, so you know which definition of the Fibonacci sequence you have.

An **arithmetic sequence**, or arithmetic progression, is a special kind of sequence in which each term has a specific quantity, called the common difference, that is added to the previous term. The common difference may be positive or negative. The general form of an arithmetic sequence containing n terms is $a_1, a_1 + d, a_1 + 2d, \dots, a_1 + (n-1)d$, where d is the common difference. The formula for the general term of an arithmetic sequence is $a_n = a_1 + (n-1)d$, where a_n is the term you are looking for and d is the common difference. To find the sum of the first n terms of an arithmetic sequence, use the formula $s_n = \frac{n}{2}(a_1 + a_n)$.

> ➤ **Review Video: <u>Arithmetic Sequence</u>**
> Visit ***mometrix.com/academy*** *and enter* ***Code*: 676885**

A **geometric sequence**, or geometric progression, is a special kind of sequence in which each term has a specific quantity, called the common ratio, multiplied by the previous term. The common ratio may be positive or negative. The general form of a geometric sequence containing n terms is $a_1, a_1 r, a_1 r^2, \dots, a_1 r^{n-1}$, where r is the common ratio. The formula for the general term of a geometric sequence is $a_n = a_1 r^{n-1}$, where a_n is the term you are looking for and r is the common ratio. To find the sum of the first n terms of a geometric sequence, use the formula $s_n = \frac{a_1(1-r^n)}{1-r}$.

Any function with the set of all natural numbers as the domain is also called a sequence. An element of a sequence is denoted by the symbol a_n, which represents the nth element of sequence a. Sequences may be arithmetic or geometric, and may be defined by a recursive definition, closed-form expression or both. Arithmetic and geometric sequences both have recursive definitions based on the first term of the sequence, as well as both having formulas to find the sum of the first n terms in the sequence, assuming you know what the first term is. The sum of all the terms in a sequence is called a **series**.

Series

An infinite series, also referred to as just a series, is a series of partial sums of a defined sequence. Each infinite sequence represents an infinite series according to the equation $\sum_{n=1}^{\infty} a_n = a_1 + a_2 + a_3 + \dots + a_n + \dots$. This notation can be shortened to $\sum_{n=1}^{\infty} a_n$ or $\sum a_n$. Every series is a sequence of partial sums, where the first partial sum is equal to the first element of the series, the second partial sum is equal to the sum of the first two elements of the series, and the nth partial sum is equal to the sum of the first n elements of the series.

Every infinite sequence of partial sums (infinite series) either converges or diverges. Like the test for convergence in a sequence, finding the limit of the sequence of partial sums will indicate whether it is a converging series or a diverging series. If there exists a real number S such that $\lim_{n \to \infty} S_n = S$, where S_n is the sequence of partial sums, then the series converges. If the limit equals infinity, then the series diverges. If $\lim_{n \to \infty} S_n = S$ and S is a real number, then S is also the convergence value of the series.

To find the sum as n approaches infinity for the sum of two convergent series, find the sum as n approaches infinity for each individual series and add the results.

$$\sum_{n=1}^{\infty} (a_n + b_n) = \sum_{n=1}^{\infty} a_n + \sum_{n=1}^{\infty} b_n$$

To find the sum as n approaches infinity for the difference between two convergent series, find the sum as n approaches infinity for each individual series and subtract the results.

$$\sum_{n=1}^{\infty} (a_n - b_n) = \sum_{n=1}^{\infty} a_n - \sum_{n=1}^{\infty} b_n$$

To find the sum as n approaches infinity for the product of a scalar and a convergent series, find the sum as n approaches infinity for the series and multiply the result by the scalar.

$$\sum_{n=1}^{\infty} k a_n = k \sum_{n=1}^{\infty} a_n$$

A **geometric series** is an infinite series in which each term is multiplied by a constant real number r, called the ratio. This is represented by the equation

$$\sum_{n=1}^{\infty} ar^{n-1} = a_1 + a_2 r + a_3 r^2 + \cdots + a_n r^{n-1} + \cdots$$

If the absolute value of r is greater than or equal to one, then the geometric series is a diverging series. If the absolute value of r is less than one but greater than zero, the geometric series is a converging series. To find the sum of a converging geometric series, use the formula

$$\sum_{n=1}^{\infty} ar^{n-1} = \frac{a}{1-r}, \text{where } 0 < |r| < 1$$

The **n^{th} term test for divergence** involves taking the limit of the nth term of a sequence and determining whether or not the limit is equal to zero. If the limit of the nth term is not equal to zero, then the series is a diverging series. This test only works to prove divergence, however. If the nth term is equal to zero, the test is inconclusive.

Cartesian Products/Relations

A Cartesian product is the product of two sets of data, X and Y, such that all elements x are a member of set X, and all elements y are a member of set Y. The product of the two sets, $X \times Y$ is the set of all ordered pairs (x, y). For example, given a standard deck of 52 playing cards, there are four possible suits (hearts, diamonds, clubs, and spades) and thirteen possible card values (the numbers 2 through 10, ace, jack, queen, and king). If the card suits are set X and the card values are set Y, then there are $4 \times 13 = 52$ possible different (x, y) combinations, as seen in the 52 cards of a standard deck.

A binary relation, also referred to as a relation, dyadic relation, or 2-place relation, is a subset of a Cartesian product. It shows the relation between one set of objects and a second set of object, or between one set of objects and itself. The prefix *bi-* means *two*, so there are always two sets involved – either two different sets, or the same set used twice. The ordered pairs of the Cartesian product are used to indicate a binary relation. Relations are possible for situations involving more than two sets, but those are not called binary relations.

The five types of relations are reflexive, symmetric, transitive, antisymmetric, and equivalence. A reflexive relation has $x\Re x$ (x related to x) for all values of x in the set. A symmetric relation has $x\Re y \Rightarrow y\Re x$ for all values of x and y in the set. A transitive relation has $(x\Re y$ and $y\Re z) \Rightarrow x\Re z$ for all values of x, y, and z in the set. An antisymmetric relation has $(x\Re y$ and $y\Re x) \Rightarrow x = y$ for all values of x and y in the set. A relation that is reflexive, symmetric, and transitive is called an equivalence relation.

Vertex-Edge Graphs

A vertex-edge graph is a set of items or objects connected by pathways or links. Below is a picture of a very basic vertex-edge graph.

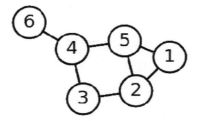

Vertex-edge graphs are useful for solving problems involving schedules, relationships, networks, or paths among a set number of objects. The number of objects may be large, but it will never be infinite. The vertices or points on the graph represent the objects and are referred to as *nodes*. The nodes are joined by line segments called *edges* or links that show the specific paths that connect the various elements represented by the nodes. The number of nodes does not have to equal the number of edges. There may be more or less, depending on the number of allowable paths.

An endpoint on a vertex-edge graph is a vertex on exactly one edge. In the case of a vertex that is an endpoint, the edge that the vertex is on is incident with the vertex. Two edges are considered to be adjacent if they share a vertex. Two vertices are considered to be adjacent if they share an edge.

In a vertex-edge graph, a loop is an edge that has the same vertex as both endpoints. To calculate the degree of a vertex in a vertex-edge graph, count the number of edges that are incident with the vertex, counting loops twice since they meet the vertex at both ends. The degree sum formula states that the sum of the degrees of all vertices on a vertex-edge graph is always equal to twice the number of edges on the graph. Thus, the sum of the degrees will never be odd, even if there are an odd number of vertices.

In a vertex-edge graph, a path is a given sequence of vertices that follows one or more edges to get from vertex to vertex. There is no jumping over spaces to get from one vertex to the next, although doubling back over an edge already traveled is allowed. A simple path is a path that does not repeat an edge in traveling from beginning to end. Think of the vertex-edge graph as a map, with the vertices as cities on the map, and the edges as roads between the cities. To get from one city to another, you must drive on the roads. A simple path allows you to complete your trip without driving on the same road twice.

In a vertex-edge graph, a circuit is a path that has the same starting and stopping point. Picturing the vertex-edge graph as a map with cities and roads, a circuit is like leaving home on vacation and then returning home after you have visited your intended destinations. You may go in one direction and then turn around, or you may go in a circle. A simple circuit on the graph completes the circuit without repeating an edge. This is like going on vacation without driving on the same road twice.

On a vertex-edge graph, any path that uses each edge exactly one time is called an Euler path. One simple way to rule out the possibility of an Euler path is to calculate the degree of each vertex. If more than two vertices have an odd degree, an Euler path is impossible. A path that uses each vertex exactly one time is called a Hamiltonian path.

If every pair of vertices is joined by an edge, the vertex-edge graph is said to be connected. If the vertex-edge graph has no simple circuits in it, then the graph is said to be a tree.

Practice Test

Practice Questions

1. Determine the number of diagonals of a dodecagon.
    a. 12
    b. 24
    c. 54
    d. 108

2. A circular bracelet contains 5 charms, A, B, C, D, and E, attached at specific points around the bracelet, with the clasp located between charms A and B. The bracelet is unclasped and stretched out into a straight line. On the resulting linear bracelet, charm C is between charms A and B, charm D is between charms A and C, and charm E is between charms C and D. Which of these statements is (are) necessarily true?

      I. The distance between charms B and E is greater than the distance between charms A and D.
      II. Charm E is between charms B and D.
      III. The distance between charms D and E is less than the distance of bracelet between charms A and C.

    a. I, II, and III
    b. II and III
    c. II only
    d. None of these is necessarily true.

3. In a town of 35,638 people, about a quarter of the population is under the age of 35. Of those, just over a third attend local K-12 schools. If the number of students in each grade is about the same, how many fourth graders likely reside in the town?
    a. Fewer than 100
    b. Between 200 and 300
    c. Between 300 and 400
    d. More than 400

4. Identical rugs are offered for sale at two local shops and one online retailer, designated Stores A, B, and C, respectively. The rug's regular sales price is $296 at Store A, $220 at Store B, and $198.00 at Store C. Stores A and B collect 8% in sales tax on any after-discount price, while Store C collects no tax but charges a $35 shipping fee. A buyer has a 30% off coupon for Store A and a $10 off coupon for Store B. Which of these lists the stores in order of lowest to highest final sales price after all discounts, taxes, and fees are applied?
    a. Store A, Store B, Store C
    b. Store B, Store C, Store A
    c. Store C, Store A, Store C
    d. Store C, Store B, Store A

5. Two companies offer monthly cell phone plans, both of which include free text messaging. Company A charges a $25 monthly fee plus five cents per minute of phone conversation, while Company B charges a $50 monthly fee and offers unlimited calling. Both companies charge the same amount when the total duration of monthly calls is
    a. 500 hours.
    b. 8 hours and 33 minutes.
    c. 8 hours and 20 minutes.
    d. 5 hours.

6. A dress is marked down by 20% and placed on a clearance rack, on which is posted a sign reading, "Take an extra 25% off already reduced merchandise." What fraction of the original price is the final sales price of the dress?
    a. $\frac{9}{20}$
    b. $\frac{11}{20}$
    c. $\frac{2}{5}$
    d. $\frac{3}{5}$

7. On a floor plan drawn at a scale of 1:100, the area of a rectangular room is 30 cm^2. What is the actual area of the room?
    a. 30,000 cm^2
    b. 3,000 cm^2
    c. 3,000 m^2
    d. 30 m^2

8. The ratio of employee wages and benefits to all other operational costs of a business is 2:3. If a business's operating expenses are $130,000 per month, how much money does the company spend on employee wages and benefits?
    a. $43,333.33
    b. $86,666.67
    c. $52,000.00
    d. $78,000.00

9. The path of ball thrown into the air is modeled by the first quadrant graph of the equation $h = -16t^2 + 64t + 5$, where h is the height of the ball in feet and t is time in seconds after the ball is thrown. What is the average rate of change in the ball's height with respect to time over the interval $[1, 3]$?
    a. 0 feet/second
    b. 48 feet/second
    c. 53 feet/second
    d. 96 feet/second

10. Zeke drove from his house to a furniture store in Atlanta and then back home along the same route. It took Zeke three hours to drive to the store. By driving an average of 20 mph faster on his return trip, Zeke was able to save an hour of diving time. What was Zeke's average driving speed on his round trip?
    a. 24 mph
    b. 48 mph
    c. 50 mph
    d. 60 mph

11. The graph below shows Aaron's distance from home at times throughout his morning run. Which of the following statements is (are) true?

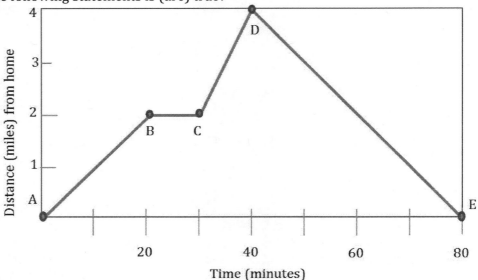

    I.   Aaron's average running speed was 6 mph.
    II. Aaron's running speed from point A to point B was the same as his running speed from point D to E.
    III. Aaron ran a total distance of four miles.

a. I only
b. II only
c. I and II
d. I, II, and III

12. Use the operation table to determine $(a * b) * (c * d)$.

$*$	a	b	c	d
a	d	a	b	c
b	a	b	c	d
c	b	c	d	a
d	c	d	a	b

a. a
b. b
c. c
d. d

13. Complete the analogy.

$$x^3 \text{ is to } \sqrt[3]{y} \text{ as } ...$$

a. $x + a$ is to $x - y$.
b. e^x is to $ln \, y, y > 0$.
c. $\frac{1}{x}$ is to $y, x, y \neq 0$.
d. $sin \, x$ is to $cos \, y$.

14. Which of these statements is (are) true for deductive reasoning?
   I. A general conclusion is drawn from specific instances.
   II. If the premises are true and proper reasoning is applied, the conclusion must be true.
a. Statement I is true
b. Statement II is true
c. Both statements are true
d. Neither statement is true

15. Given that premises "all a are b," "all b are d," and "no b are c" are true and that premise "all b are e" is false, determine the validity and soundness of the following arguments:
   Argument I: All a are b. No b are c. Therefore, no a are c.
   Argument II: All a are b. All d are b. Therefore, all d are a.
   Argument III: All a are b. All b are e. Therefore, all a are e.

a.

	Invalid	Valid	Sound
Argument I		X	X
Argument II	X		
Argument III		X	

b.

	Invalid	Valid	Sound
Argument I	X		
Argument II		X	X
Argument III	X		

c.

	Invalid	Valid	Sound
Argument I		X	X
Argument II		X	X
Argument III	X		

d.

	Invalid	Valid	Sound
Argument I		X	X
Argument II	X		
Argument III	X		

16. If $p \rightarrow q$ is true, which of these is also necessarily true?

    a. $q \rightarrow p$

    b. $\sim p \rightarrow \sim q$

    c. $\sim q \rightarrow \sim p$

    d. None of these

17. Given statements p and q, which of the following is the truth table for the statement $q \leftrightarrow \sim(p \wedge q)$?

a.

p	q	$q \leftrightarrow \sim(p \wedge q)$
T	T	F
T	F	T
F	T	T
F	F	T

b.

p	q	$q \leftrightarrow \sim(p \wedge q)$
T	T	T
T	F	T
F	T	T
F	F	F

c.

p	q	$q \leftrightarrow \sim(p \wedge q)$
T	T	F
T	F	F
F	T	F
F	F	T

d.

p	q	$q \leftrightarrow \sim(p \wedge q)$
T	T	F
T	F	F
F	T	T
F	F	F

18. Which of the following is the truth table for logic circuit shown below?

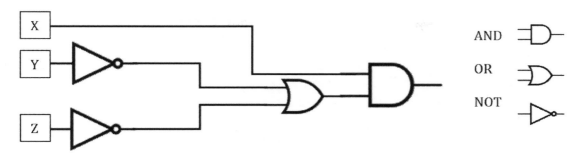

a.

X	Y	Z	Output
0	0	0	1
0	0	1	0
0	1	0	0
0	1	1	0
1	0	0	0
1	0	1	0
1	1	0	0
1	1	1	1

c.

X	Y	Z	Output
0	0	0	0
0	0	1	0
0	1	0	0
0	1	1	1
1	0	0	1
1	0	1	1
1	1	0	1
1	1	1	0

b.

X	Y	Z	Output
0	0	0	0
0	0	1	1
0	1	0	1
0	1	1	1
1	0	0	1
1	0	1	1
1	1	0	1
1	1	1	1

d.

X	Y	Z	Output
0	0	0	0
0	0	1	0
0	1	0	0
0	1	1	0
1	0	0	1
1	0	1	1
1	1	0	1
1	1	1	0

19. Which of these is a major contribution of the Babylonian civilization to the historical development of mathematics?
    a. The division of an hour into 60 minutes, and a minute into 60 seconds, and a circle into 360 degrees
    b. The development of algebra as a discipline separate from geometry
    c. The use of deductive reasoning in geometric proofs
    d. The introduction of Boolean logic and algebra

20. Which mathematician is responsible for what is often called the most remarkable and beautiful mathematical formula, $e^{i\pi} + 1 = 0$?
    a. Pythagoras
    b. Euclid
    c. Euler
    d. Fermat

21. Which of these demonstrates the relationship between the sets of prime numbers, real numbers, natural numbers, complex numbers, rational numbers, and integers?
$\mathbb{P}$ – Prime; $\mathbb{R}$ – Real; $\mathbb{N}$ – Natural; $\mathbb{C}$ – Complex; $\mathbb{Q}$ – Rational; $\mathbb{Z}$ – Integer
    a. $\mathbb{P} \subseteq \mathbb{Q} \subseteq \mathbb{R} \subseteq \mathbb{Z} \subseteq \mathbb{C} \subseteq \mathbb{N}$
    b. $\mathbb{P} \subseteq \mathbb{N} \subseteq \mathbb{Z} \subseteq \mathbb{Q} \subseteq \mathbb{R} \subseteq \mathbb{C}$
    c. $\mathbb{C} \subseteq \mathbb{R} \subseteq \mathbb{Q} \subseteq \mathbb{Z} \subseteq \mathbb{N} \subseteq \mathbb{P}$
    d. None of these

22. To which of the following sets of numbers does -4 **NOT** belong?
    a. The set of whole numbers
    b. The set of rational numbers
    c. The set of integers
    d. The set of real numbers

23. Which of these forms a group?
    a. The set of prime numbers under addition
    b. The set of negative integers under multiplication
    c. The set of negative integers under addition
    d. The set of non-zero rational numbers under multiplication

24. Simplify $\frac{2+3i}{4-2i}$.
    a. $\frac{1}{10} + \frac{4}{5}i$
    b. $\frac{1}{10}$
    c. $\frac{7}{6} + \frac{2}{3}i$
    d. $\frac{1}{10} + \frac{3}{10}i$

25. Simplify $|(2 - 3i)^2 - (1 - 4i)|$.
    a. $\sqrt{61}$
    b. $-6 - 8i$
    c. $6 + 8i$
    d. 10

26. Which of these sets forms a group under multiplication?
    a. $\{-i, 0, i\}$
    b. $\{-1, 1, i, -i\}$
    c. $\{i, 1\}$
    d. $\{ i, -i, 1\}$

27. The set $\{a, b, c, d\}$ forms a group under operation #. Which of these statements is (are) true about the group?

#	a	b	c	d
a	c	d	b	a
b	d	c	a	b
c	b	a	d	c
d	a	b	c	d

    I.   The identity element of the group is d.
    II.   The inverse of c is c.
    III.   The operation # is commutative.

a. I
b. III
c. I, III
d. I, II, III

28. If the square of twice the sum of x and three is equal to the product of twenty-four and x, which of these is a possible value of x?
a. $6 + 3\sqrt{2}$
b. $\frac{3}{2}$
c. $-3i$
d. -3

29. Given that x is a prime number and that the greatest common factor of x and y is greater than 1, compare the two quantities.

Quantity A	Quantity B
y	the least common multiple of x and y

a. Quantity A is greater.
b. Quantity B is greater.
c. The two quantities are the same.
d. The relationship cannot be determined from the given information.

30. If a, b, and c are even integers and $3a^2 + 9b^3 = c$, which of these is the largest number which must be factor of c?
a. 2
b. 3
c. 6
d. 12

31. Which of these relationships represents y as a function of x?
   a. $x = y^2$

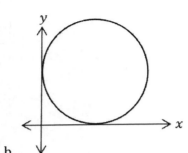

   b.

   c. $y = [\![x]\!]$s

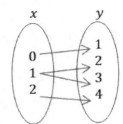

   d.

32. Express the area of the given triangle as a function of x.

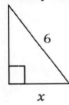

   a. $A(x) = 3x$
   b. $A(x) = \dfrac{x\sqrt{36-x^2}}{2}$
   c. $A(x) = \dfrac{x^2}{2}$
   d. $A(x) = 18 - \dfrac{x^2}{2}$

33. Find $[g \circ f]x$ when $f(x) = 2x + 4$ and $g(x) = x^2 - 3x + 2$.
   a. $4x^2 + 10x + 6$
   b. $2x^2 - 6x + 8$
   c $4x^2 + 13x + 18.$
   d. $2x^2 - 3x + 6$

34. Given the partial table of values for $f(x)$ and $g(x)$, find $f(g(-4))$. (Assume that $f(x)$ and $g(x)$ are the simplest polynomials that fit the data.)

x	f(x)	g(x)
-2	8	1
-1	2	3
0	0	5
1	2	7
2	8	9

    a. 69
    b. 31
    c. 18
    d. –3

35. If $f(x)$ and $g(x)$ are inverse functions, which of these is the value of x when $f(g(x)) = 4$?
    a. –4
    b. $\dfrac{1}{4}$
    c. 2
    d. 4

36. Determine which pair of equations are **NOT** inverses.
    a. $y = x + 6; y = x - 6$
    b. $y = 2x + 3; y = 2x - 3$
    c. $y = \dfrac{2x+3}{x-1}; y = \dfrac{x+3}{x-2}$
    d. $y = \dfrac{x-1}{2}; y = 2x + 1$

37. Which of these statements is (are) true for function $g(x)$?

$$g(x) = \begin{cases} 2x - 1 & x \geq 2 \\ -x + 3 & x < 2 \end{cases}$$

    I. $g(3) = 0$
    II.   The graph of $g(x)$ is discontinuous at $x = 2$.
    III.   The range of $g(x)$ is all real numbers.
    a. II
    b. III
    c. I, II
    d. II, III

38. Which of the following piecewise functions can describe the graph below?

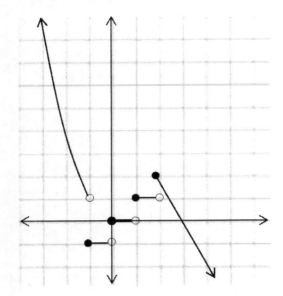

a. $f(x) = \begin{cases} x^2 & x < -1 \\ [\![x]\!] & -1 \le x < 2 \\ -2x + 6 & x \ge 2 \end{cases}$

b. $f(x) = \begin{cases} x^2 & x \le -1 \\ [\![x]\!] & -1 \le x \le 2 \\ -2x + 6 & x > 2 \end{cases}$

c. $f(x) = \begin{cases} (x + 1)^2 & x < -1 \\ [\![x]\!] + 1 & -1 \le x < 2 \\ -2x + 6 & x \ge 2 \end{cases}$

d. $f(x) = \begin{cases} (x + 1)^2 & x < -1 \\ [\![x - 1]\!] & -1 \le x < 2 \\ -2x + 6 & x \ge 2 \end{cases}$

39. Which of the following could be the graph of $y = a(x + b)(x + c)^2$ if $a > 0$?

a.

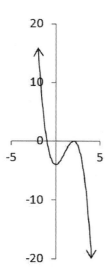

c.

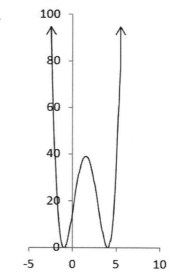

b.

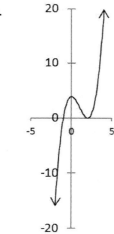

d.

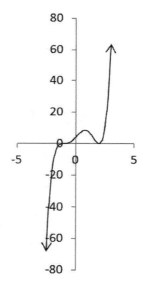

40. A school is selling tickets to its production of *Annie Get Your Gun*. Student tickets cost $3 each, and non-student tickets are $5 each. In order to offset the costs of the production, the school must earn at least $300 in ticket sales. Which graph shows the number of tickets the school must sell to offset production costs?

a.

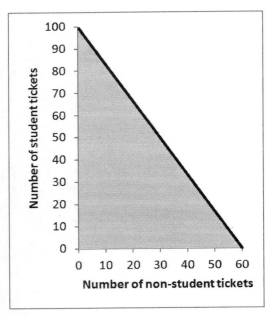

c.

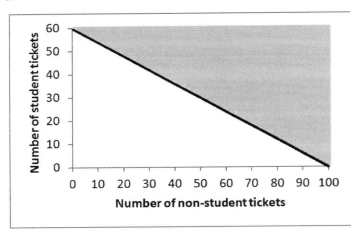

b.

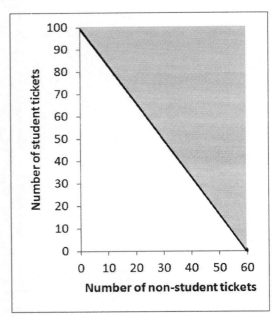

d.

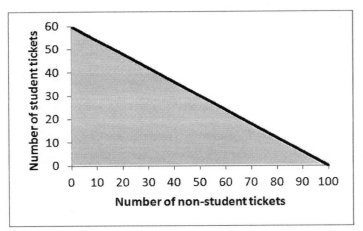

41. Which of these is the equation graphed below?
    a. $y = -2x^2 - 4x + 1$
    b. $y = -x^2 - 2x + 5$
    c. $y = -x^2 - 2x + 2$
    d. $y = -\frac{1}{2}x^2 - x + \frac{5}{2}$

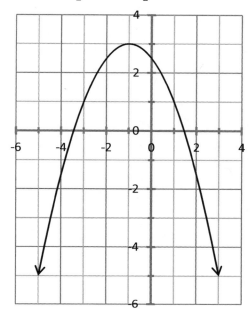

42. Solve $7x^2 + 6x = -2$.
    a. $x = \frac{-3 \pm \sqrt{23}}{7}$
    b. $x = \pm i\sqrt{5}$
    c. $x = \pm \frac{2i\sqrt{2}}{7}$
    d. $x = \frac{-3 \pm i\sqrt{5}}{7}$

43. Solve the system of equations.
$$3x + 4y = 2$$
$$2x + 6y = -2$$

    a. $\left(0, \frac{1}{2}\right)$
    b. $\left(\frac{2}{5}, \frac{1}{5}\right)$
    c. $(2, -1)$
    d. $\left(-1, \frac{5}{4}\right)$

44. Which system of linear inequalities has no solution?
   a. $x - y < 3$
   $x - y \geq -3$
   b. $y \leq 6 - 2x$
   $\frac{1}{3}y + \frac{2}{3}x \geq 2$
   c. $6x + 2y \leq 12$
   $3x \geq 8 - y$
   d. $x + 4y \leq -8$
   $y + 4x > -8$

45. The cost of admission to a theme park is shown below.

Under age 10	Ages 10-55	Over age 65
$15	$25	$20

Yesterday, the theme park sold 810 tickets and earned $14,500. There were twice as many children under 10 at the park as there were other visitors. If x, y, and z represent the number of $15, $25, and $20 tickets sold, respectively, which of the following matrix equations can be used to find the number of each type of ticket sold?

a. $\begin{bmatrix} 1 & 1 & 1 \\ 15 & 25 & 20 \\ 1 & -2 & -2 \end{bmatrix} \begin{bmatrix} x \\ y \\ z \end{bmatrix} = \begin{bmatrix} 810 \\ 14500 \\ 0 \end{bmatrix}$

b. $\begin{bmatrix} 1 & 1 & 1 \\ 15 & 25 & 20 \\ 1 & -2 & -2 \end{bmatrix} \begin{bmatrix} 810 \\ 14500 \\ 0 \end{bmatrix} = \begin{bmatrix} x \\ y \\ z \end{bmatrix}$

c. $\begin{bmatrix} 1 & 15 & 1 \\ 1 & 25 & -2 \\ 1 & 20 & -2 \end{bmatrix} \begin{bmatrix} x \\ y \\ z \end{bmatrix} = \begin{bmatrix} 810 \\ 14500 \\ 0 \end{bmatrix}$

d. $\begin{bmatrix} 1 & 15 & 1 \\ 1 & 25 & -2 \\ 1 & 20 & -2 \end{bmatrix} \begin{bmatrix} 810 \\ 14500 \\ 0 \end{bmatrix} = \begin{bmatrix} x \\ y \\ z \end{bmatrix}$

46. Solve the system of equations.
$$2x - 4y + z = 10$$
$$-3x + 2y - 4z = -7$$
$$x + y - 3z = -1$$

   a. $(-1, -3, 0)$
   b. $(1, -2, 0)$
   c. $(-\frac{3}{4}, -\frac{21}{8}, -1)$
   d. No solution

47. Solve $x^4 + 64 = 20x^2$.
   a. $x = \{2, 4\}$
   b. $x = \{-2, 2, -4, 4\}$
   c. $x = \{2i, 4i\}$
   d. $x = \{-2i, 2i, -4i, 4i\}$

48. Solve $3x^3y^2 - 45x^2y = 15x^3y - 9x^2y^2$ for x and y.
    a. $x = \{0, -3\},\ y = \{0, 5\}$
    b. $x = \{0\},\ y = \{0\}$
    c. $x = \{0, -3\},\ y = \{0\}$
    d. $x = \{0\},\ y = \{0, 5\}$

49. Which of these statements is true for functions $f(x)$, $g(x)$, and $h(x)$?
$$f(x) = 2x - 2$$
$$g(x) = 2x^2 - 2$$
$$h(x) = 2x^3 - 2$$
    a. The degree of each polynomial function is 2.
    b. The leading coefficient of each function is –2.
    c. Each function has exactly one real zero at $x = 1$.
    d. None of these is true for functions $f(x)$, $g(x)$, and $h(x)$.

50. Which of these can be modeled by a quadratic function?
    a. The path of a sound wave
    b. The path of a bullet
    c. The distance an object travels over time when the rate is constant
    d. Radioactive decay

51. Which of these is equivalent to $\log_y 256$ if $2\log_4 y + \log_4 16 = 3$?
    a. 16
    b. 8
    c. 4
    d. 2

52. Simplify $\dfrac{(x^2y)(2xy^{-2})^3}{16x^5y^2} + \dfrac{3}{xy}$

    a. $\dfrac{3x + 24y^6}{8xy^7}$

    b. $\dfrac{x + 6y^6}{2xy^7}$

    c. $\dfrac{x + 24y^5}{8xy^6}$

    d. $\dfrac{x + 6y^5}{2xy^6}$

53. Given: $f(x) = 10^x$. If $f(x) = 5$, which of these approximates x?
    a. 100,000
    b. 0.00001
    c. 0.7
    d. 1.6

54. Which of these could be the equation of the function graphed below?
    a. $f(x) = x^2$
    b. $f(x) = \sqrt{x}$
    c. $f(x) = 2^x$
    d. $f(x) = log_2 x$

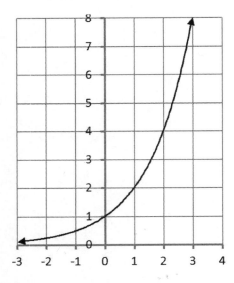

55. Which of these statements is **NOT** necessarily true when $f(x) = log_b x$ and $b > 1$?
    a. The x-intercept of the graph of $f(x)$ is 1.
    b. The graph of $f(x)$ passes through $(b, 1)$
    c. $f(x) < 0$ when $x < 1$
    d. If $g(x) = b^x$, the graph of $f(x)$ is symmetric to the graph of $g(x)$ with respect to $y = x$.

56. A colony of *Escherichia coli* is inoculated from a Petri dish into a test tube containing 50 mL of nutrient broth. The test tube is placed in a 37°C incubator/shaker; after one hour, the number of bacteria in the test tube is determined to be 8×10^6. Given that the doubling time of *E. coli* is 20 minutes with agitation at 37°C, approximately how many bacteria should the test tube contain after eight hours of growth?
    a. 2.56×10^8
    b. 2.05×10^9
    c. 1.7×10^{10}
    d. 1.7×10^{13}

57. The strength of an aqueous acid solution is measured by pH. $pH = -log[H^+]$, where $[H^+]$ is the molar concentration of hydronium ions in the solution. A solution is acidic if its pH is less than 7. The lower the pH, the stronger the acid; for example, gastric acid, which has a pH of about 1, is a much stronger acid than urine, which has a pH of about 6. How many times stronger is an acid with a pH of 3 than an acid with pH of 5?
    a. 2
    b. 20
    c. 100
    d. 1000

58. Simplify $\sqrt{\dfrac{-28x^6}{27y^5}}$.

    a. $\dfrac{2x^3 i\sqrt{21y}}{9y^3}$

    b. $\dfrac{2x^3 i\sqrt{21y}}{27y^4}$

    c. $\dfrac{-2x^3\sqrt{21y}}{9y^3}$

    d. $\dfrac{12x^3 yi\sqrt{7}}{27y^2}$

59. Which of these does **NOT** have a solution set of $\{x: -1 \leq x \leq 1\}$?

    a. $-4 \leq 2 + 3(x - 1) \leq 2$

    b. $-2x^2 + 2 \geq x^2 - 1$

    c. $\dfrac{11 - |3x|}{7} \geq 2$

    d. $3|2x| + 4 \leq 10$

60. Solve $2 - \sqrt{x} = \sqrt{x - 20}$.

    a. $x = 6$

    b. $x = 36$

    c. $x = 144$

    d. No solution

61. Solve $\dfrac{x-2}{x-1} = \dfrac{x-1}{x+1} + \dfrac{2}{x-1}$.

    a. $x = 2$

    b. $x = -5$

    c. $x = 1$

    d. No solution

62. Which of these equations is represented by the graph below?

a. $y = \dfrac{3}{x^2 - x - 2}$

b. $y = \dfrac{3x + 3}{x^2 - x - 2}$

c. $y = \dfrac{1}{x+1} + \dfrac{1}{x-2}$

d. None of these

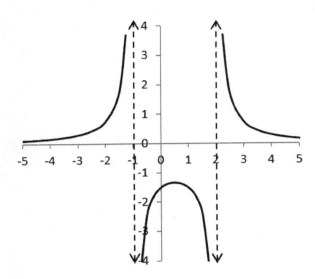

63. Which of the graphs shown represents $f(x) = -2|-x + 4| - 1$?

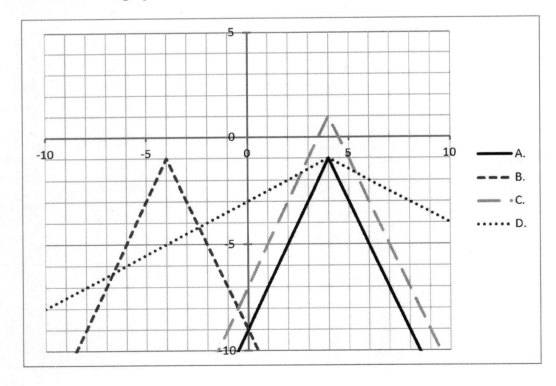

64. Which of these functions includes 1 as an element of the domain and 2 as an element of the range?

a. $y = \frac{1}{x-1} + 1$

b. $y = -\sqrt{x+2} - 1$

c. $y = |x+2| - 3$

d. $y = \begin{cases} x & x < 1 \\ -x - 3 & x \geq 1 \end{cases}$

65. Which of the following statements is (are) true when $f(x) = \frac{x^2-x-6}{x^3+2x^2-x-2}$?

    I. The graph $f(x)$ has vertical asymptotes at $x = -2$, $x = -1$, and $x = 1$.
    II. The x- and y-intercepts of the graph of $f(x)$ are both 3.

a. I
b. II
c. I and II
d. Neither statement is true.

66. In the 1600s, Galileo Galilei studied the motion of pendulums and discovered that the period of a pendulum, the time it takes to complete one full swing, is a function of the square root of the length of its string: $2\pi\sqrt{\frac{L}{g}}$, where L is the length of the string and g is the acceleration due to gravity.

Consider two pendulums released from the same pivot point and at the same angle, $\theta = 30°$. Pendulum 1 has a mass of 100 g, while Pendulum 2 has a mass of 200 g. If Pendulum 1 has a period four times the period of Pendulum 2, what is true of the lengths of the pendulums' strings?

a. The length of Pendulum 1's string is four times the length of Pendulum 2's string.
b. The length of Pendulum 1's string is eight times the length of Pendulum 2's string.
c. The length of Pendulum 1's string is sixteen times the length of Pendulum 2's string.
d. The length of Pendulum 1's string is less than the length of Pendulum 2's string.

67. At today's visit to her doctor, Josephine was prescribed a liquid medication with instructions to take 25 cc's every four hours. She filled the prescription on her way to work, but when it came time to take the medicine, she realized that the pharmacist did not include a measuring cup. Josephine estimated that the plastic spoon in her desk drawer was about the same size as a teaspoon and decided to use it to measure the approximate dosage. She recalled that one cubic centimeter (cc) is equal to one milliliter (mL) but was not sure how many milliliters were in a teaspoon. So, she noted that a two-liter bottle of soda contains about the same amount as a half-gallon container of milk and applied her knowledge of the customary system of measurement to determine how many teaspoons of medicine to take. Which of these calculations might she have used to approximate her dosage?

a. $25 \cdot \dfrac{1}{1000} \cdot \dfrac{2}{0.5} \cdot 16 \cdot 48$

b. $25 \cdot \dfrac{1}{100} \cdot \dfrac{0.5}{2} \cdot 16 \cdot 4 \cdot 12$

c. $\dfrac{1000}{25} \cdot \dfrac{0.5}{2} \cdot 16 \cdot 4 \cdot 12$

d. $\dfrac{25}{1000} \cdot \dfrac{1}{4} \cdot 16 \cdot 48$

68. If 1" on a map represents 60 ft, how many yards apart are two points if the distance between the points on the map is 10"?

a. 1800
b. 600
c. 200
d. 2

69. Roxana walks x meters west and $x + 20$ meters south to get to her friend's house. On a neighborhood map which has a scale of 1cm:10 m, the distance between Roxana's house and her friend's house is 10 cm. How far did Roxana walk to her friend's house?

a. 60 m
b. 80 m
c. 100 m
d. 140 m

70. For $\triangle ABC$, what is AB?

a. 3
b. 10
c. 12
d. 15

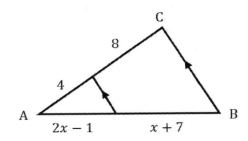

71. To test the accuracy and precision of two scales, a student repeatedly measured the mass of a 10 g standard and recorded these results.

	Trial 1	Trial 2	Trial 3	Trial 4
Scale 1	9.99 g	9.98 g	10.02g	10.01g
Scale 2	10.206 g	10.209 g	10.210 g	10.208 g

Which of these conclusions about the scales is true?
   a. Scale 1 has an average percent error of 0.15%, and Scale 2 has an average percent error of 2.08%. Scale 1 is more accurate and precise than Scale 2.
   b. Scale 1 has an average percent error of 0.15%, and Scale 2 has an average percent error of 2.08%. Scale 1 is more accurate than Scale 2; however, Scale 2 is more precise.
   c. Scale 1 has an average percent error of 0%, and Scale 2 has an average percent error of 2.08%. Scale 1 is more accurate and precise than Scale 2.
   d. Scale 1 has an average percent error of 0%, and Scale 2 has an average percent error of 2.08%. Scale 1 is more accurate than Scale 2; however, Scale 2 is more precise.

72. A developer decides to build a fence around a neighborhood park, which is positioned on a rectangular lot. Rather than fencing along the lot line, he fences x feet from each of the lot's boundaries. By fencing a rectangular space 141 yd² smaller than the lot, the developer saves $432 in fencing materials, which cost $12 per linear foot. How much does he spend?
   a. $160
   b. $456
   c. $3,168
   d. The answer cannot be determined from the given information.

73. Natasha designs a square pyramidal tent for her children. Each of the sides of the square base measures x ft, and the tent's height is h feet. If Natasha were to increase by 1 ft the length of each side of the base, how much more interior space would the tent have?
   a. $\frac{h(x^2+2x+1)}{3}$ ft³
   b. $\frac{h(2x+1)}{3}$ ft³
   c. $\frac{x^2h+3}{3}$ ft³
   d. 1 ft³

74. A rainbow pattern is designed from semi-circles as shown below.

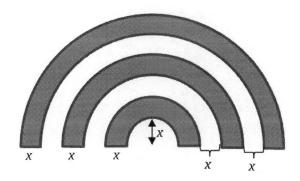

Which of the following gives the area A of the shaded region as a function of x?

a. $A = \frac{21x^2\pi}{2}$

b. $A = 21x^2\pi$

c. $A = 42x^2\pi$

d. $A = 82x^2\pi$

75. Categorize the following statements as axioms of Euclidean, hyperbolic, or elliptical geometry.

      I. In a plane, for any line l and point A not on l, no lines which pass through A intersect l.

      II. In a plane, for any line l and point A not on l, exactly one line which passes through A does not intersect l.

      III. In a plane, for any line l and point A not on l, all lines which pass through A intersect l.

a.

Statement I	Elliptical geometry
Statement II	Euclidean geometry
Statement III	Hyperbolic geometry

b.

Statement I	Hyperbolic geometry
Statement II	Euclidean geometry
Statement III	Elliptical geometry

c.

Statement I	Hyperbolic geometry
Statement II	Elliptical geometry
Statement III	Euclidean geometry

d.

Statement I	Elliptical geometry
Statement II	Hyperbolic geometry
Statement III	Euclidean geometry

76. As shown below, four congruent isosceles trapezoids are positioned such that they form an arch. Find x for the indicated angle.

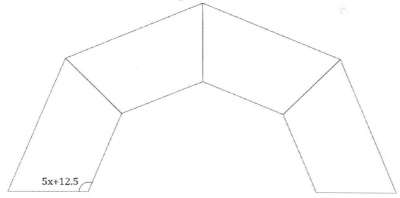

5x+12.5

    a. $x = 11$
    b. $x = 20$
    c. $x = 24.5$
    d. The value of x cannot be determined from the information given.

77. A circle is inscribed inside quadrilateral $ABCD$. $\overline{CD}$ is bisected by the point at which it is tangent to the circle. If $AB = 14, BC = 10, DC = 8$, then

    a. $AD = 11$
    b. $AD = 2\sqrt{34}$
    c. $AD = 12$
    d. $AD = 17.5$

78. Which of the following equations gives the area A of the triangle below as a function of a and b?

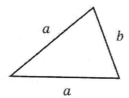

    a. $\dfrac{2a^2 - b^2}{4}$
    b. $\dfrac{ab - a^2}{2}$
    c. $\dfrac{b\sqrt{a^2 - b^2}}{2}$
    d. $\dfrac{b\sqrt{4a^2 - b^2}}{4}$

79. Given the figure and the following information, find DE to the nearest tenth.

$\overline{AD}$ is an altitude of $\triangle ABC$
$\overline{DE}$ is an altitude of triangle $\triangle ADC$
$BD \cong DC$
$BC = 24$; $AD = 5$

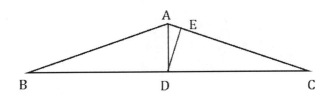

a. 4.2
b. 4.6
c. 4.9
d. 5.4

80. A cube inscribed in a sphere has a volume of 64 cubic units. What is the volume of the sphere in cubic units?
    a. $4\pi\sqrt{3}$
    b. $8\pi\sqrt{3}$
    c. $32\pi\sqrt{3}$
    d. $256\pi\sqrt{3}$

Questions 81 and 82 are based on the following proof:

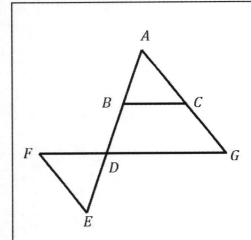

Statement	Reason
1. $\overline{BC}\|\overline{FG}$	Given
2.	
3. $\overline{FD} \cong \overline{BC}$	Given
4. $\overline{AB} \cong \overline{DE}$	Given
5. $\triangle ABC \cong \triangle EDF$	____81.____
6. ____82.____	
7. $\overline{FE}\|\overline{AG}$	

Given: $\overline{BC}\|\overline{FG}$; $\overline{FD} \cong \overline{BC}$; $\overline{AB} \cong \overline{DE}$
Prove: $\overline{FE}\|\overline{AG}$

81. Which of the following justifies step 5 in the proof?
    a. AAS
    b. SSS
    c. ASA
    d. SAS

82. Step 6 in the proof should contain which of the following statements?
   a. $\angle BAC \cong \angle DEF$
   b. $\angle ABC \cong \angle EDF$
   c. $\angle ACB \cong \angle EFD$
   d. $\angle GDA \cong \angle EDF$

83. Which of these is **NOT** a net of a cube?

a.
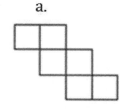

b.
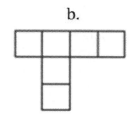

c.
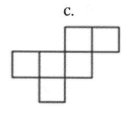

d.
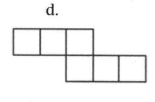

84. Identify the cross-section polygon formed by a plane containing the given points on the cube.

   a. Rectangle
   b. Trapezoid
   c. Pentagon
   d. Hexagon

85. Which of these represents the equation of a sphere which is centered in the xyz-space at the point (1, 0, -2) and which has a volume of 36π cubic units?
   a. $x^2 + y^2 + z^2 - 2x + 4z = 4$
   b. $x^2 + y^2 + z^2 + 2x - 4z = 4$
   c. $x^2 + y^2 + z^2 - 2x + 4z = -2$
   d. $x^2 + y^2 + z^2 + 2x - 4z = 2$

86. A triangle has vertices (0,0,0), (0,0,4), and (0,3,0) in the xyz-space. In cubic units, what is the difference in the volume of the solid formed by rotating the triangle about the z-axis and the solid formed by rotating the triangle about the y-axis?
   a. 0
   b. 4π
   c. 5π
   d. 25

87. If the midpoint of a line segment graphed on the xy-coordinate plane is $(3, -1)$ and the slope of the line segment is -2, which of these is a possible endpoint of the line segment?
   a. $(-1,1)$
   b. $(0, -5)$
   c. $(7,1)$
   d. $(5, -5)$

88. The vertices of a polygon are $(2,3)$, $(8,1)$, $(6,-5)$, and $(0,-3)$. Which of the following describes the polygon most specifically?
 a. Parallelogram
 b. Rhombus
 c. Rectangle
 d. Square

89. What is the radius of the circle defined by the equation $x^2 + y^2 - 10x + 8y + 29 = 0$?
 a. $2\sqrt{3}$
 b. $2\sqrt{5}$
 c. $\sqrt{29}$
 d. 12

90. Which of these describes the graph of the equation $2x^2 - 3y^2 - 12x + 6y - 15 = 0$?
 a. Circular
 b. Elliptical
 c. Parabolic
 d. Hyperbolic

91. The graph of $f(x)$ is a parabola with a focus of (a, b) and a directrix of $y = -b$, and $g(x)$ represents a transformation of $f(x)$. If the vertex of the graph of $g(x)$ is $(a, 0)$, which of these is a possible equation for $g(x)$ for nonzero integers a and b?
 a. $g(x) = f(x) + b$
 b. $g(x) = -f(x)$
 c. $g(x) = f(x + a)$
 d. $g(x) = f(x - a) + b$

92. A triangle with vertices $A(-4,2)$, $B(-1,3)$, and $C(-5,7)$ is reflected across $y = x + 2$ to give $\Delta A'B'C'$, which is subsequently reflected across the y-axis to give $\Delta A''B''C''$. Which of these statements is true?
 a. A 90° rotation of ΔABC about $(-2,0)$ gives $\Delta A''B''C''$.
 b. A reflection of ΔABC about the x-axis gives $\Delta A''B''C''$.
 c. A 270° rotation of ΔABC about $(0,2)$ gives $\Delta A''B''C''$.
 d. A translation of ΔABC two units down gives $\Delta A''B''C''$.

93. For which of these does a rotation of 120° about the center of the polygon map the polygon onto itself?
 a. Square
 b. Regular hexagon
 c. Regular octagon
 d. Regular decagon

94. Line segment $\overline{PQ}$ has endpoints (a, b) and (c, b). If $\overline{P'Q'}$ is the translation of $\overline{PQ}$ along a diagonal line such that P' is located at point (c, d), what is the area of quadrilateral $PP'Q'Q$?
 a. $|a - c| \cdot |b - d|$
 b. $|a - b| \cdot |c - d|$
 c. $|a - d| \cdot |b - c|$
 d. $(a - c)^2$

95. For the right triangle below, which of the following is a true statement of equality?

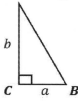

a. $\tan B = \dfrac{a}{b}$

b. $\cos B = \dfrac{a\sqrt{a^2+b^2}}{a^2+b^2}$

c. $\sec B = \dfrac{\sqrt{a^2+b^2}}{b}$

d. $\csc B = \dfrac{a^2+b^2}{b}$

96. A man looks out of a window of a tall building at a 45° angle of depression and sees his car in the parking lot. When he turns his gaze downward to a 60° angle of depression, he sees his wife's car. If his car is parked 60 feet from his wife's car, about how far from the building did his wife park her car?

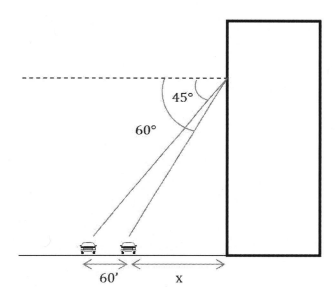

a. 163 feet

b. 122 feet

c. 82 feet

d. 60 feet

97. What is the exact value of $\tan\left(-\dfrac{2\pi}{3}\right)$?

a. $\sqrt{3}$

b. $-\sqrt{3}$

c. $\dfrac{\sqrt{3}}{3}$

d. 1

98. If $\sin \theta = \frac{1}{2}$ when $\frac{\pi}{2} < \theta < \pi$, what is the value of θ?

    a. $\frac{\pi}{6}$

    b. $\frac{\pi}{3}$

    c. $\frac{2\pi}{3}$

    d. $\frac{5\pi}{6}$

99. Which of the following expressions is equal to $\cos \theta \cot \theta$?

    a. $sin\ \theta$
    b. $sec\ \theta\ tan\ \theta$
    c. $csc\ \theta - sin\ \theta$
    d. $sec\ \theta - sin\ \theta$

100. Solve $\sec^2\theta = 2 \tan \theta$ for $0 < \theta \le 2\pi$.

    a. $\theta = \frac{\pi}{6}$ or $\frac{7\pi}{6}$

    b. $\theta = \frac{\pi}{4}$ or $\frac{5\pi}{4}$

    c. $\theta = \frac{3\pi}{4}$ or $\frac{7\pi}{4}$

    d. There is no solution to the equation.

101. A car is driving along the highway at a constant speed when it runs over a pebble, which becomes lodged in one of the tire's treads. If this graph represent the height h of the pebble above the road in inches as a function of time t in seconds, which of these statements is true?

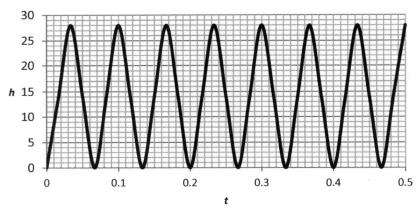

    a. The outer radius of the tire is 14 inches, and the tire rotates 900 times per minute.
    b. The outer radius of the tire is 28 inches, and the tire rotates 900 times per minute.
    c. The outer radius of the tire is 14 inches, and the tire rotates 120 times per minute.
    d. The outer radius of the tire is 28 inches, and the tire rotates 120 times per minute.

Below are graphed functions $f(x) = a_1 \sin(b_1 x)$ and $g(x) = a_2 \cos(b_2 x)$; a_1 and a_2 are integers, and b_1 and b_2 are positive rational numbers. Use this information to answer questions 102-103:

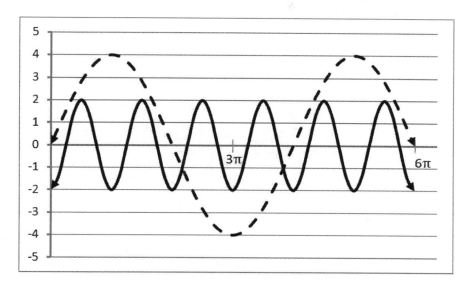

102. Which of the following statements is true?
    a. The graph of $f(x)$ is represented by a solid line.
    b. The amplitude of the graph of $g(x)$ is 4.
    c. $0 < b_1 < 1$.
    d. $b_2 = \pi$.

103. Which of the following statements is true?
    a. $0 < a_2 < a_1$
    b. $a_2 < 0 < a_1$
    c. $0 < a_1 < a_2$
    d. $a_2 < a_1 < 0$

104. A weight suspended on a spring is at its equilibrium point five inches above the top of a table. When the weight is pulled down two inches, it bounces above the equilibrium point and returns to the point from which it was released in one second. Which of these can be used to model the weight's height h above the table as a function of time t in seconds?
    a. $h = -2\cos(2\pi t) + 5$
    b. $h = 5\sin(t) - 2$
    c. $h = -2\sin(2\pi t) + 5$
    d. $h = -2\cos(0.5\pi t) + 3$

105. Evaluate $\lim_{x \to -3} \frac{x^3 + 3x^2 - x - 3}{x^2 - 9}$.
    a. 0
    b. $\frac{1}{3}$
    c. $-\frac{4}{3}$
    d. ∞

106. Evaluate $\lim_{x\to\infty}\frac{x^2+2x-3}{2x^2+1}$.

   a. 0

   b. $\frac{1}{2}$

   c. -3

   d. ∞

107. Evaluate $\lim_{x\to3^+}\frac{|x-3|}{3-x}$.

   a. 0

   b. –1

   c. 1

   d. ∞

108. If $f(x)=\frac{1}{4}x^2-3$, find the slope of the line tangent to graph of $f(x)$ at $x=2$.

   a. –2

   b. 0

   c. 1

   d. 4

109. If $f(x)=2x^3-3x^2+4$, what is $\lim_{h\to0}\frac{f(2+h)-f(2)}{h}$?

   a. -4

   b. 4

   c. 8

   d. 12

110. Find the derivative of $f(x)=e^{3x^2-1}$.

   a. $6xe^{6x}$

   b. e^{3x^2-1}

   c. $(3x^2-1)e^{3x^2-2}$

   d. $6xe^{3x^2-1}$

111. Find the derivative of $f(x)=\ln(2x+1)$.

   a. $\frac{1}{2x+1}$

   b. $2e^{2x+1}$

   c. $\frac{2}{2x+1}$

   d. $\frac{1}{2}$

112. For functions $f(x)$, $g(x)$, and $h(x)$, determine the limit of the function as x approaches 2 and the continuity of the function at $x = 2$.

a.

$\lim_{x \to 2+} f(x) = 4$ $\lim_{x \to 2-} f(x) = 2$ $f(2) = 2$	$lim_{x \to 2} f(x)$ DNE	The function $f(x)$ is discontinuous at 2.
$\lim_{x \to 2+} g(x) = 2$ $\lim_{x \to 2-} g(x) = 2$ $g(2) = 4$	$\lim_{x \to 2} g(x) = 2$	The function $g(x)$ is discontinuous at 2.
$\lim_{x \to 2+} h(x) = 2$ $\lim_{x \to 2-} h(x) = 2$ $h(2) = 2$	$\lim_{x \to 2} h(x) = 2$	The function $h(x)$ is continuous at 2.

b.

$\lim_{x \to 2+} f(x) = 4$ $\lim_{x \to 2-} f(x) = 2$ $f(2) = 2$	$lim_{x \to 2} f(x)$ DNE	The function $f(x)$ is continuous at 2.
$\lim_{x \to 2+} g(x) = 2$ $\lim_{x \to 2-} g(x) = 2$ $g(2) = 4$	$lim_{x \to 2} g(x)$ DNE	The function $g(x)$ is continuous at 2.
$\lim_{x \to 2+} h(x) = 2$ $\lim_{x \to 2-} h(x) = 2$ $h(2) = 2$	$\lim_{x \to 2} h(x) = 2$	The function $h(x)$ is continuous at 2.

c.

$\lim_{x \to 2+} f(x) = 4$ $\lim_{x \to 2-} f(x) = 2$ $f(2) = 2$	$\lim_{x \to 2} f(x) = 2$	The function $f(x)$ is continuous at 2.
$\lim_{x \to 2+} g(x) = 2$ $\lim_{x \to 2-} g(x) = 2$ $g(2) = 4$	$\lim_{x \to 2} g(x) = 2$	The function $g(x)$ is discontinuous at 2.
$\lim_{x \to 2+} h(x) = 2$ $\lim_{x \to 2-} h(x) = 2$ $h(2) = 2$	$\lim_{x \to 2} h(x) = 2$	The function $h(x)$ is continuous at 2.

d.

$\lim_{x \to 2+} f(x) = 4$ $\lim_{x \to 2-} f(x) = 2$ $f(2) = 2$	$\lim_{x \to 2} f(x) = 2$	The function $f(x)$ is discontinuous at 2.
$\lim_{x \to 2+} g(x) = 2$ $\lim_{x \to 2-} g(x) = 2$ $g(2) = 4$	$\lim_{x \to 2} g(x) = 2$	The function $g(x)$ is discontinuous at 2.
$\lim_{x \to 2+} h(x) = 2$ $\lim_{x \to 2-} h(x) = 2$ $h(2) = 2$	$\lim_{x \to 2} h(x) = 2$	The function $h(x)$ is continuous at 2.

113. Find $f''(x)$ if $f(x) = 2x^4 - 4x^3 + 2x^2 - x + 1$.
   a. $24x^2 - 24x + 4$
   b. $8x^3 - 12x^2 + 4x - 1$
   c. $32x^2 - 36x^2 + 8$
   d. $\frac{2}{5}x^5 - x^4 + \frac{2}{3}x^3 - \frac{1}{2}x^2 + x + c$

114. If $f(x) = 4x^3 - x^2 - 4x + 2$, which of the following statements is(are) true of its graph?
   I. The point $\left(-\frac{1}{2}, 3\frac{1}{4}\right)$ is a relative maximum.
   II. The graph of f is concave upward on the interval $\left(-\infty, \frac{1}{2}\right)$.
   a. I
   b. II
   c. I and II
   d. Neither I nor II

115. Suppose the path of a baseball hit straight up from three feet above the ground is modeled by the first quadrant graph of the function $h = -16t^2 + 50t + 3$, where t is the flight time of the ball in seconds and h is the height of the ball in feet. What is the velocity of the ball two seconds after it is hit?
   a. 39 ft/s upward
   b. 19.5 ft/s upward
   c. 19.5 ft/s downward
   d. 14 ft/s downward

116. A manufacturer wishes to produce a cylindrical can which can hold up to 0.5 L of liquid. To the nearest tenth, what is the radius of the can which requires the least amount of material to make?
   a. 2.8 cm
   b. 4.3 cm
   c. 5.0 cm
   d. 9.2 cm

117. Approximate the area A under the curve by using a Riemann sum with $\Delta x = 1$.

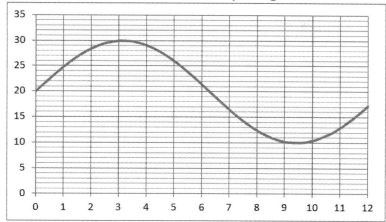

   a. $209 < A < 211$
   b. $230 < A < 235$
   c. $238 < A < 241$
   d. $246 < A < 250$

118. To the nearest hundredth, what is the area in square units under the curve of $f(x) = \frac{1}{x}$ on [1,2]?
   a. 0.50
   b. 0.69
   c. 1.30
   d. 1.50

119. Calculate $\int 3x^2 + 2x - 1 \; dx$.
   a. $x^3 + x^2 - x + c$
   b. $6x^2 + 2$
   c. $\frac{3}{2}x^3 + 2x^2 - x + c$
   d. $6x^2 + 2 + c$

120. Calculate $\int 3x^2 e^{x^3} \; dx$
   a. $x^3 e^{x^3} + c$
   b. $e^{x^3} + c$
   c. $x^3 e^{\frac{x^4}{4}} + c$
   d. $\ln x^3 + c$

121. Find the area A of the finite region between the graphs of $y = -x + 2$ and $y = x^2 - 4$.
   a. 18
   b. $\frac{125}{6}$
   c. $\frac{45}{2}$
   d. 25

122. The velocity of a car which starts at position 0 at time 0 is given by the equation $v(t) = 12t - t^2$ for $0 \le t \le 12$. Find the position of the car when its acceleration is 0.

    a. 18
    b. 36
    c. 144
    d. 288

123. Which of these graphs is **NOT** representative of the data set shown below?

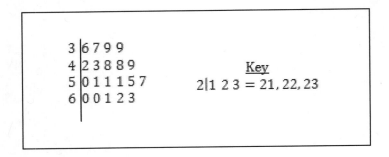

```
3 | 6 7 9 9
4 | 2 3 8 8 9                Key
5 | 0 1 1 1 5 7        2|1 2 3 = 21, 22, 23
6 | 0 0 1 2 3
```

a.

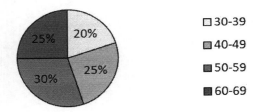

b.

Frequency

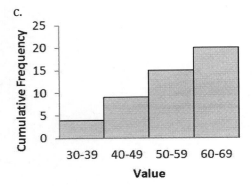

    □ 30-39
    ▨ 40-49
    ■ 50-59
    ■ 60-69

c.

d. All of these graphs represent the data set.

124. Which of these would best illustrate change over time?
  a. Pie chart
  b. Line graph
  c. Box-and-whisker plot
  d. Venn diagram

125. Which of these is the least biased sampling technique?
  a. To assess his effectiveness in the classroom, a teacher distributes a teacher evaluation to all of his students. Responses are anonymous and voluntary.
  b. To determine the average intelligence quotient (IQ) of students in her school of 2,000 students, a principal uses a random number generator to select 300 students by student identification number and has them participate in a standardized IQ test.
  c. To determine which video game is most popular among his fellow eleventh graders at school, a student surveys all of the students in his English class.
  d. Sixty percent of students at the school have a parent who is a member of the Parent-Teacher Association (PTA). To determine parent opinions regarding school improvement programs, the Parent-Teacher Association (PTA) requires submission of a survey response with membership dues.

126. Which of these tables properly displays the measures of central tendency which can be used for nominal, interval, and ordinal data?

a.

	Mean	Median	Mode
Nominal			x
Interval	x	x	x
Ordinal		x	x

b.

	Mean	Median	Mode
Nominal			x
Interval	x	x	x
Ordinal	x	x	x

c.

	Mean	Median	Mode
Nominal	x	x	x
Interval	x	x	x
Ordinal	x	x	x

d.

	Mean	Median	Mode
Nominal			x
Interval	x	x	
Ordinal	x	x	x

Use the following data to answer questions 127-129:

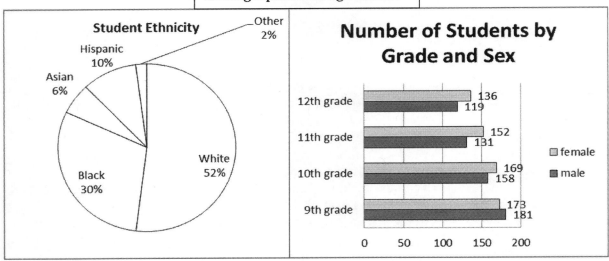

127. Which of these is the greatest quantity?
    a. The average number of male students in the 11th and 12th grades
    b. The number of Hispanic students at the school
    c. The difference in the number of male and female students at the school
    d. The difference in the number of 9th and 12th grader students at the school

128. Compare the two quantities.

Quantity A	Quantity B
The percentage of white students at the school, rounded to the nearest whole number	The percentage of female students at the school, rounded to the nearest whole number

    a. Quantity A is greater.
    b. Quantity B is greater.
    c. The two quantities are the same.
    d. The relationship cannot be determined from the given information.

129. An eleventh grader is chosen at random to represent the school at a conference. What is the approximate probability that the student is male?
    a. 0.03
    b. 0.11
    c. 0.22
    d. 0.46

The box-and-whisker plot displays student test scores by class period. Use the data to answer questions 130 through 132:

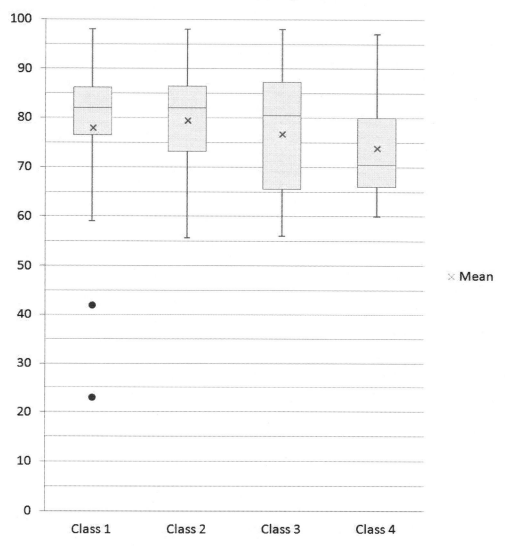

30. Which class has the greatest range of test scores?
    a. Class 1
    b. Class 2
    c. Class 3
    d. Class 4

131. What is the probability that a student chosen at random from class 2 made above a 73 on this test?
    a. 0.25
    b. 0.5
    c. 0.6
    d. 0.75

132. Which of the following statements is true of the data?
    a. The mean better reflects student performance in class 1 than the median.
    b. The mean test score for class 1 and 2 is the same.
    c. The median test score for class 1 and 2 is the same.
    d. The median test score is above the mean for class 4.

Copyright © Mometrix Media. You have been licensed one copy of this document for personal use only. Any other reproduction or redistribution is strictly prohibited. All rights reserved.

133. In order to analyze the real estate market for two different zip codes within the city, a realtor examines the most recent 100 home sales in each zip code. She considered a house which sold within the first month of its listing to have a market time of one month; likewise, she considered a house to have a market time of two months if it sold after having been on the market for one month but by the end of the second month. Using this definition of market time, she determined the frequency of sales by number of months on the market. The results are displayed below.

Which of the following is a true statement for these data?

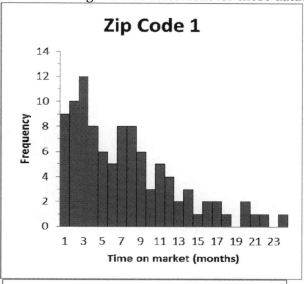

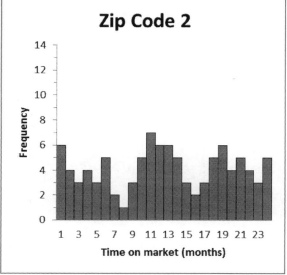

a. The median time a house spends on the market in Zip Code 1 is five months less than Zip Code 2

b. On average, a house spent seven months longer on the market in Zip Code 2 than in Zip Code 1.

c. The mode time on the market is higher for Zip Code 1 than for Zip Code 2.

d. The median time on the market is less than the mean time on the market for Zip Code 1.

134. Attending a summer camp are 12 six-year-olds, 15 seven-year-olds, 14 eight-year-olds, 12 nine-year-olds, and 10 ten-year-olds. If a camper is randomly selected to participate in a special event, what is the probability that he or she is at least eight years old?

    a. $\frac{2}{9}$

    b. $\frac{22}{63}$

    c. $\frac{4}{7}$

    d. $\frac{3}{7}$

135. A small company is divided into three departments as shown. Two individuals are chosen at random to attend a conference. What is the approximate probability that two women from the same department will be chosen?

	Department 1	Department 2	Department 3
Women	12	28	16
Men	18	14	15

    a. 8.6%
    b. 10.7%
    c. 11.2%
    d. 13.8%

136. A random sample of 90 students at an elementary school were asked these three questions:

Do you like carrots?
Do you like broccoli?
Do you like cauliflower?

The results of the survey are shown below. If these data are representative of the population of students at the school, which of these is most probable?

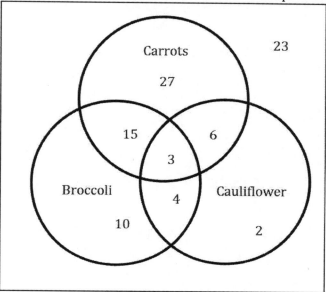

    a. A student chosen at random likes broccoli.
    b. If a student chosen at random likes carrots, he also likes at least one other vegetable.
    c. If a student chosen at random likes cauliflower and broccoli, he also likes carrots.
    d. A student chosen at random does not like carrots, broccoli, or cauliflower.

Use the information below to answer questions 137 and 138:

Each day for 100 days, a student tossed a single misshapen coin three times in succession and recorded the number of times the coin landed on heads. The results of his experiment are shown below.

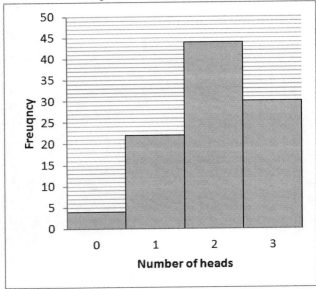

137. Given these experimental data, which of these approximates P(heads) for a single flip of this coin.

a. 0.22
b. 0.5
c. 0.67
d. 0.74

138. Which of these shows the graphs of the probability distributions from ten flips of this misshapen coin and ten flips of a fair coin?

a.

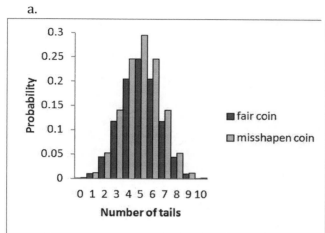

c.

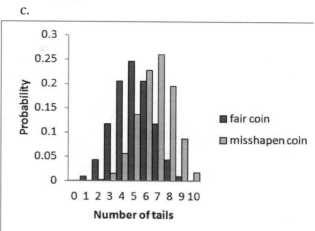

b.

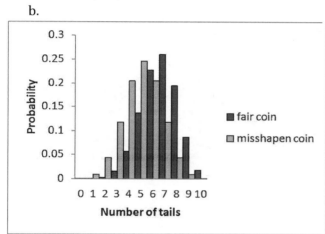

d.

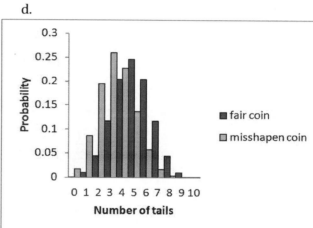

139. Which of these does **NOT** simulate randomly selecting a student from a group of 11 students?
a. Assigning each student a unique card value of A, 1, 2, 3, 4, 5, 6, 7, 8, 9, or J, removing queens and kings from a standard deck of 52 cards, shuffling the remaining cards, and drawing a single card from the deck
b. Assigning each student a unique number 0-10 and using a computer to randomly generate a number within that range
c. Assigning each student a unique number from 2 to 12 ; rolling two dice and finding the sum of the numbers on the dice
d. All of these can be used as a simulation of the event.

140. Gene P has three possible alleles, or gene forms, called a, b and c. Each individual carries two copies of Gene P, one of which is inherited from his or her mother and the other of which is inherited from his or her father. If the two copies of Gene P are of the same form, the individual is homozygous for that allele; otherwise, the individual is heterozygous. A simulation is performed to determine the genotypes, or genetic make-ups, of 500 individuals selected at random from the population. 500 two-digit numbers are generator using a random number generator. Based on the relative frequencies of each allele, the digit 0 is assigned to represent allele a, the digits 1 and 2 to represent allele b, and the digits 3-9 to represent allele c.

```
28 93 97 37 92 00 27 21 87 13 62 62 15 31 55 09 47 07 54 88 38 88 10
98 34 01 45 14 34 46 38 61 93 22 37 39 57 03 93 50 53 16 28 65 81 60
21 12 13 10 19 91 04 18 49 01 99 30 11 16 00 48 04 63 59 24 02 42 23
06 32 52 19 18 94 94 46 63 87 41 79 39 85 20 43 20 15 03 39 33 77 45
66 77 70 92 25 27 68 71 89 35 98 55 85 47 60 97 12 92 53 44 45 41 51
22 09 23 81 33 04 35 43 48 32 80 36 95 64 56 34 74 55 37 64 84 51 50
25 99 51 94 19 46 10 44 17 25 75 52 47 35 70 65 08 50 98 09 02 24 30
59 00 03 21 40 30 86 16 53 91 28 17 97 58 75 76 73 83 54 40 54 13 38
36 67 74 80 63 12 41 27 96 61 66 05 60 69 96 15 56 82 57 31 83 26 24
78 42 76 49 56 06 57 78 67 02 96 40 82 29 14 07 29 62 90 31 08 26 71
61 18 22 84 23 33 49 29 90 07 08 05 14 59 72 86 44 69 68 99 06 11 95
43 72 58 28 93 97 37 92 00 27 21 87 13 62 62 15 31 55 09 47 07 54 88
38 88 10 98 34 01 45 14 34 46 38 61 93 22 37 39 57 03 93 50 53 16 28
65 81 60 21 12 13 10 19 91 04 18 49 01 99 30 11 16 00 48 04 63 59 24
02 42 23 06 32 52 19 18 94 94 46 63 87 41 79 39 85 20 43 20 15 03 39
33 77 45 66 77 70 92 25 27 68 71 89 35 98 55 85 47 60 97 12 92 53 44
45 41 51 22 09 23 81 33 04 35 43 48 32 80 36 95 64 56 34 74 55 37 64
84 51 50 25 99 51 94 19 46 10 44 17 25 75 52 47 35 70 65 08 50 98 09
02 24 30 59 00 03 21 40 30 86 16 53 91 28 17 97 58 75 76 73 83 54 40
54 13 38 36 67 74 80 63 12 41 27 96 61 66 05 60 69 96 15 56 82 57 31
83 26 24 78 42 76 49 56 06 57 78 67 02 96 40 82 29 14 07 29 62 90 31
08 26 71 61 18 22 84 23 33 49 29 90 07 08 05 14 59
```

Using the experimental probability that an individual will be homozygous for allele a (light grey) or for allele b (dark grey), predict the number of individuals in a population of 100,000 who will be homozygous for either allele.
  a. 2,800
  b. 5,000
  c. 5,400
  d. 9,000

141. The intelligence quotients (IQs) of a randomly selected group of 300 people are normally distributed with a mean IQ of 100 and a standard deviation of 15. In a normal distribution, approximately 68% of values are within one standard deviation of the mean. About how many individuals from the selected group have IQs of at least 85?
  a. 96
  b. 200
  c. 216
  d. 252

142. How many different seven-digit telephone numbers can be created in which no digit repeats and in which zero cannot be the first digit?
    a. 5,040
    b. 35,280
    c. 544,320
    d. 3, 265,920

143. A teacher wishes to divide her class of twenty students into four groups, each of which will have three boys and two girls. How many possible groups can she form?
    a. 248
    b. 6,160
    c. 73,920
    d. 95,040

144. In how many distinguishable ways can a family of five be seated at a circular table with five chairs if Tasha and Mac must be kept separated?
    a. 6
    b. 12
    c. 24
    d. 60

145. Which of these defines the recursive sequence $a_1 = -1, a_{n+1} = a_n + 2$ explicitly?
    a. $a_n = 2n - 3$
    b. $a_n = -n + 2$
    c. $a_n = n - 2$
    d. $a_n = -2n + 3$

146. What is the sum of the series 200 + 100 + 50 + 25 + ...?
    a. 300
    b. 400
    c. 600
    d. The sum is infinite.

147. For vector $v = (4, 3)$ and vector $w = (-3,4)$, find $2(v + w)$.
    a. $(2, 14)$
    b. $(14, -2)$
    c. $(1,7)$
    d. $(7, -1)$

148. Simplify $\begin{bmatrix} 2 & 0 & -5 \end{bmatrix} \left(\begin{bmatrix} 4 \\ 2 \\ -1 \end{bmatrix} - \begin{bmatrix} 3 \\ 5 \\ -5 \end{bmatrix} \right)$.

    a. $[-18]$

    b. $\begin{bmatrix} 2 \\ 0 \\ -20 \end{bmatrix}$

    c. $\begin{bmatrix} 2 & 0 & -20 \end{bmatrix}$

    d. $\begin{bmatrix} 2 & 0 & -5 \\ -6 & 0 & 15 \\ 8 & 0 & -20 \end{bmatrix}$

149. Consider three sets, of which one contains the set of even integers, one contains the factors of twelve, and one contains elements 1, 2, 4, and 9. If each set is assigned the name A, B, or C, and $A \cap B \subseteq B \cap C$, which of these must be set C?

    a. The set of even integers
    b. The set of factors of 12
    c. The set {1, 2, 4, 9}
    d. The answer cannot be determined from the given information.

150. Last year, Jenny tutored students in math, in chemistry, and for the ACT. She tutored ten students in math, eight students in chemistry, and seven students for the ACT. She tutored five students in both math and chemistry, and she tutored four students both in chemistry and for the ACT, and five students both in math and for the ACT. She tutored three students in all three subjects. How many students did Jenny tutor last year?

    a. 34
    b. 25
    c. 23
    d. 14

Answers and Explanations

1. C: Because drawing a dodecagon and counting its diagonals is an arduous task, it is useful to employ a different problem-solving strategy. One such strategy is to draw polygons with fewer sides and look for a pattern in the number of the polygons' diagonals.

(triangle)	3	0
(square)	4	2
(pentagon)	5	5
(hexagon)	6	9
Heptagon	7	14
Octagon	8	20

A quadrilateral has two more diagonals than a triangle, a pentagon has three more diagonals than a quadrilateral, and a hexagon has four more diagonals than a pentagon. Continue this pattern to find that a dodecagon has 54 diagonals.

2. B: The problem does not give any information about the size of the bracelet or the spacing between any of the charms. Nevertheless, creating a simple illustration which shows the order of the charms will help when approaching this problem. For example, the circle below represents the bracelet, and the dotted line between A and B represents the clasp. On the right, the line shows the stretched out bracelet and possible positions of charms C, D, and E based on the parameters.

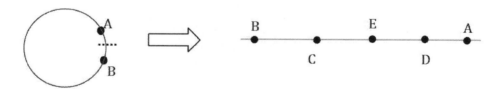

From the drawing above, it appears that statement I is true, but it is not necessarily so. The alternative drawing below also shows the charms ordered correctly, but the distance between B and E is now less than that between D and A.

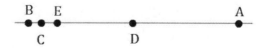

Statement II must be true: charm E must lie between B and D. Statement III must also be true: the distance between charms E and D must be less than that between C and A, which includes charms E and D in the space between them.

3. B: The population is approximately 36,000, so one quarter of the population consists of about 9,000 individuals under age 35. A third of 9,000 is 3,000, the approximate number of students in grades K-12. Since there are thirteen grades, there are about 230 students in each grade. So, the number of fourth graders is between 200 and 300.

4. A: The final sales price of the rug is $1.08(0.7 \cdot \$296) = \$223.78$ at Store A, $1.08(\$220 - \$10) = \$226.80$ at Store B, and $\$198 + \$35 = \$233$ at Store C.

5. C: The expression representing the monthly charge for Company A is $\$25 + \$0.05m$, where m is the time in minutes spent talking on the phone. Set this expression equal to the monthly charge for Company B, which is $50. Solve for m to find the number of minutes for which the two companies charge the same amount:

$$\$25 + \$0.05m = \$50$$
$$\$0.05m = \$25$$
$$m = 500$$

Notice that the answer choices are given in hours, not in minutes. Since there are 60 minutes in an hour, $m = \frac{500}{60}$ hours $= 8\frac{1}{3}$ hours. One-third of an hour is twenty minutes, so $m = 8$ hours, 20 minutes.

6. D: When the dress is marked down by 20%, the cost of the dress is 80% of its original price; thus, the reduced price of the dress can be written as $\frac{80}{100}x$, or $\frac{4}{5}x$, where x is the original price. When discounted an extra 25%, the dress costs 75% of the reduced price, or $\frac{75}{100}\left(\frac{4}{5}x\right)$, or $\frac{3}{4}\left(\frac{4}{5}x\right)$, which simplifies to $\frac{3}{5}x$. So the final price of the dress is three-fifths of the original price.

7. D: Since there are 100 cm in a meter, on a 1:100 scale drawing, each centimeter represents one meter. Therefore, an area of one square centimeter on the drawing represents one square meter in actuality. Since the area of the room in the scale drawing is 30 cm^2, the room's actual area is 30 m^2.

Another way to determine the area of the room is to write and solve an equation, such as this one:
$\frac{l}{100} \cdot \frac{w}{100} = 30$ cm^2, where l and w are the dimensions of the actual room

$$\frac{lw}{1000} = 30 \text{ cm}^2$$
$$lw = 300{,}000 \text{ cm}^2$$
$$\text{Area} = 300{,}000 \text{ cm}^2$$

Since this is not one of the answer choices, convert cm^2 to m^2: $300{,}000$ cm$^2 \cdot \frac{1 \text{ m}}{100 \text{ cm}} \cdot \frac{1 \text{ m}}{100 \text{ cm}} = 30$ m^2.

8. C: Since the ratio of wages and benefits to other costs is 2:3, the amount of money spent on wages and benefits is $\frac{2}{5}$ of the business's total expenditure. $\frac{2}{5} \cdot \$130{,}000 = \$52{,}000$.

9. A: The height of the ball is a function of time, so the equation can be expressed as $f(t) = -16t^2 + 64t + 5$, and the average rate of change can be found by calculating $\frac{f(3)-f(1)}{3-1}$.

$$\frac{-16(3)^2 + 64(3) + 5 - [-16(1)^2 + 64(1) + 5]}{2} = \frac{-144 + 192 + 5 - (-16 + 64 + 5)}{2} = \frac{0}{2} = 0$$

Alternatively, the rate of change can be determined by finding the slope of the secant line through points $(1, f(1))$ and $(3, f(3))$. Notice that this is a horizontal line, which has a slope of 0.

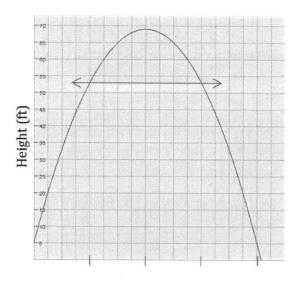

Time (sec)

10. B: Since rate in mph $= \frac{\text{distance in miles}}{\text{time in hours}}$, Zeke's driving speed on the way to Atlanta and home from Atlanta in mph can be expressed as d/3 and d/2, respectively, when d=distance between Zeke's house and his destination . Since Zeke drove 20 mph faster on his way home, $\frac{d}{2} - \frac{d}{3} = 20$.

$$6\left(\frac{d}{2} - \frac{d}{3} = 20\right)$$
$$3d - 2d = 120$$
$$d = 120$$

Since the distance between Zeke's house and the store in Atlanta is 120 miles, Zeke drove a total distance of 240 miles in five hours. Therefore, his average speed was $\frac{240 \text{ miles}}{5 \text{ hours}} = 48$ mph.

11. C: Aaron ran four miles from home and then back again, so he ran a total of eight miles. Therefore, statement III is false. Statements I and II, however, are both true. Since Aaron ran eight miles in eighty minutes, he ran an average of one mile every ten minutes, or six miles per hour; he ran two miles from point A to B in 20 minutes and four miles from D to E in 40 minutes, so his running speed between both sets of points was the same.

12. D: First, use the table to determine the values of $(a * b)$ and $(c * d)$.

*	a	b	c	d
a	d	a	b	c
b	a	b	c	d
c	b	c	d	a
d	c	d	a	b

$(a * b) = a$ and $(c * d) = a$, so $(a * b) * (c * d) = a * a$, which is equal to d.

13. B: When $y = x^3$, $x = \sqrt[3]{y}$. Similarly, when $y = e^x$, $x = \ln y$ for $y > 0$. On the other hand, when $y = x + a$, $x = y - a$; when $y = 1/x$, $x = 1/y$ for $x, y \neq 0$; and when $y = \sin x$, $x = \sin^{-1} y$.

14. B: Deductive reasoning moves from one or more general statements to a specific, while inductive reasoning makes a general conclusion based on a series of specific instances or observations. Whenever the premises used in deductive reasoning are true, the conclusion drawn is necessarily true. In inductive reasoning, it is possible for the premises to be true and the conclusion to be false since there may exist an exception to the general conclusion drawn from the observations made.

15. A: The first argument's reasoning is valid, and since its premises are true, the argument is also sound. The second argument's reasoning is invalid; that the premises are true is irrelevant. (For example, consider the true premises "all cats are mammals" and "all dogs are mammals;" it cannot be logically concluded that all dogs are cats.) The third argument's reasoning is valid, but since one of its premises is false, the argument is not sound.

16. C: The logical representation $p \rightarrow q$ means that p implies q. In other words, if p, then q. Unlike the contrapositive (Choice C), neither the converse (choice A) nor the inverse (choice B) is necessarily true. For example, consider this statement: all cats are mammals. This can be written as an if/then statement: if an animal is a cat, then the animal is a mammal. The converse would read, "If an animal is a mammal, then the animal is a cat;" of course, this is not necessarily true since there are many mammals other than cats. The inverse statement, "If an animal is not a cat, then the animal is not a mammal," is false. The contrapositive, "If an animal is not a mammal, then the animal is not a cat" is true since there are no cats which are not mammals.

17. D: The symbol $\wedge$ is the logical conjunction symbol. In order for statement $(p \wedge q)$ to be true, both statements p and q must be true. The $\sim$ symbol means "not," so if $(p \wedge q)$ is true, then $\sim(p \wedge q)$ is false, and if $(p \wedge q)$ is false, then $\sim(p \wedge q)$ is true. The statement $q \leftrightarrow \sim(p \wedge q)$ is true when the value of q is the same as the value of $\sim(p \wedge q)$.

p	q	$(p \wedge q)$	$\sim(p \wedge q)$	$q \leftrightarrow \sim(p \wedge q)$
T	T	T	F	F
T	F	F	T	F
F	T	F	T	T
F	F	F	T	F

18. D: The value "0" means "false," and the value "1" means "true." For the logical disjunction "or," the output value is true if either or both input values are true, else it is false. For the logical conjunction "and," the output value is true only if both input values are true. "Not A" is true when A is false and is false when A is true.

X	Y	Z	not Y	not Z	not Y or not Z	X and (not Y or not Z)
0	0	0	1	1	1	0
0	0	1	1	0	1	0
0	1	0	0	1	1	0
0	1	1	0	0	0	0
1	0	0	1	1	1	1
1	0	1	1	0	1	1
1	1	0	0	1	1	1
1	1	1	0	0	0	0

19. A: The Babylonians used a base-60 numeral system, which is still used in the division of an hour into 60 minutes, a minute into 60 seconds, and a circle into 360 degrees. (The word "algebra" and its development as a discipline separate from geometry are attributed to the Arabic/Islamic civilization. The Greek philosopher Thales is credited with using deductive reasoning to prove geometric concepts. Boolean logic and algebra was introduced by British mathematician George Boole.)

20. C: Leonhard Euler made many important contributions to the field of mathematics. One such contribution, Euler's formula $e^{i\varphi} = \cos\varphi + i\sin\varphi = 0$, can be written as $e^{i\pi} + 1 = 0$ when $\varphi = \pi$. This identity is considered both mathematically remarkable and beautiful, as it links together five important mathematical constants, $e, i, \pi, 0$ and 1.

21. B: The notation $\mathbb{P} \subseteq \mathbb{N} \subseteq \mathbb{Z} \subseteq \mathbb{Q} \subseteq \mathbb{R} \subseteq \mathbb{C}$ means that the set of prime numbers is a subset of the set natural numbers, which is a subset of the set of integers, which is a subset of the set of rational numbers, which is a subset of the set real numbers, which is a subset of the set of complex numbers.

22. A: The set of whole numbers, $\{0, 1, 2, 3, \ldots\}$, does not contain the number -4. Since -4 is an integer, it is also a rational number and a real number.

23. D: In order for a set to be a group under operation $*$,
1. The set must be closed under that operation. In other words, when the operation is performed on any two members of the set, the result must also be a member of that set.
2. The set must demonstrate associativity under the operation: $a * (b * c) = (a * b) * c$
3. There must exist an identity element e in the group: $a * e = e * a = a$
4. For every element in the group, there must exist an inverse element in the group: $a * b = b * a = e$

Note: the group need not be commutative for every pair of elements in the group. If the group demonstrates commutativity, it is called an abelian group.

The set of prime numbers under addition is not closed. For example, 3+5=8, and 8 is not a member of the set of prime numbers. Similarly, the set of negative integers under multiplication is not closed since the product of two negative integers is a positive integer. Though the set of negative integers under addition is closed and is associative, there exists no identity element (the number zero in this case) in the group. The set of positive rational numbers under multiplication is closed and associative; the multiplicative identity 1 is a member of the group, and for each element in the group, there is a multiplicative inverse (reciprocal).

24. A: First, multiply the numerator and denominator by the denominator's conjugate, $4 + 2i$. Then, simplify the result and write the answer in the form $a + bi$.

$$\frac{2 + 3i}{4 - 2i} \cdot \frac{4 + 2i}{4 + 2i} = \frac{8 + 4i + 12i + 6i^2}{16 - 4i^2} = \frac{8 + 16i - 6}{16 + 4} = \frac{2 + 16i}{20} = \frac{1}{10} + \frac{4}{5}i$$

25. D: First, simplify the expression within the absolute value symbol.
$$|(2 - 3i)^2 - (1 - 4i)|$$
$$|4 - 12i + 9i^2 - 1 + 4i|$$
$$|4 - 12i - 9 - 1 + 4i|$$
$$|-6 - 8i|$$
The absolute value of a complex number is its distance from 0 on the complex plane. Use the Pythagorean Theorem (or the 3-4-5 Pythagorean triple and similarity) to find the distance of $-6 - 8i$ from the origin.

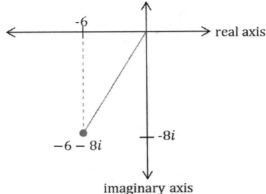

Since the distance from the origin to the point $-6 - 8i$ is 10, $|-6 - 8i|$=10.

26. B: In order for a set to be a group under operation $*$,
1. The set must be closed under that operation. In other words, when the operation is performed on any two members of the set, the result must also be a member of that set.
2. The set must demonstrate associativity under the operation: $a * (b * c) = (a * b) * c$
3. There must exist an identity element e in the group: $a * e = e * a = a$
4. For every element in the group, there must exist an inverse element in the group: $a * b = b * a = e$

Choice A can easily be eliminated as the correct answer because the set $\{-i, 0, i\}$ does not contain the multiplicative identity 1. Though choices C and D contain the element 1, neither is closed: for example, since $i \cdot i = -1$, -1 must be an element of the group. Choice B is closed, contains the multiplicative identity 1, and the inverse of each element is included in the set as well. Of course, multiplication is an associative operation, so the set $\{-1, 1, i, -i\}$ forms a group under multiplication

×	-1	1	i	$-i$
-1	1	-1	$-i$	i
1	-1	1	i	$-i$
i	$-i$	i	-1	1
$-i$	i	$-i$	1	-1

- 138 -

27. D: The identity element is d since $d\#a = a\#d = a$, $d\#b = b\#d = b$, $d\#c = c\#d = c$, and $d\#d = d$. The inverse of element c is c since $c\#c = d$, the identity element. The operation # is commutative because $a\#b = b\#a$, $a\#c = c\#a$, etc. Rather than check that the operation is commutative for each pair of elements, note that elements in the table display symmetry about the diagonal elements; this indicates that the operation is indeed commutative.

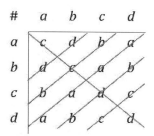

28. C: "The square of twice the sum of x and three is equal to the product of twenty-four and x" is represented by the equation $[2(x + 3)]^2 = 24x$. Solve for x.
$$[2(x + 3)]^2 = 24x$$
$$[2x + 6]^2 = 24x$$
$$4x^2 + 24x + 36 = 24x$$
$$4x^2 = -36$$
$$x^2 = -9$$
$$x = \pm\sqrt{-9}$$
$$x = \pm 3i$$

So, $-3i$ is a possible value of x.

29. C: If x is a prime number and that the greatest common factor of x and y is greater than 1, the greatest common factor of x and y must be x. The least common multiple of two numbers is equal to the product of those numbers divided by their greatest common factor. So, the least common multiple of x and y is $\frac{xy}{x} = y$. Therefore, the values in the two columns are the same.

30. D: Since a and b are even integers, each can be expressed as the product of 2 and an integer. So, if we write $a = 2x$ and $b = 2y$, $3(2x)^2 + 9(2y)^3 = c$.
$$3(4x^2) + 9(8y^3) = c$$
$$12x^2 + 72y^3 = c$$
$$12(x^2 + 6y^3) = c$$

Since c is the product of 12 and some other integer, 12 must be a factor of c. Incidentally, the numbers 2, 3, and 6 must also be factors of c since each is also a factor of 12.

31. C: Choice C is the equation for the greatest integer function. A function is a relationship in which for every element of the domain (x), there is exactly one element of the range (y). Graphically, a relationship between x and y can be identified as a function if the graph passes the vertical line test.

The first relation is a parabola on its side, which fails the vertical line test for functions. A circle (Choice B) also fails the vertical line test and is therefore not a function. The relation in Choice D pairs two elements of the range with one of the elements of the domain, so it is also not a function.

32. B: The area of a triangle is $A = \frac{1}{2}bh$, where b and h are the lengths of the triangle's base and height, respectively. The base of the given triangle is x, but the height is not given. Since the triangle is a right triangle and the hypotenuse is given, the triangle's height can be found using the Pythagorean Theorem.

$$x^2 + h^2 = 6^2$$
$$h = \sqrt{36 - x^2}$$

To find the area of the triangle in terms of x, substitute $\sqrt{36 - x^2}$ for the height and x for the base of the triangle into the area formula.

$$A = \frac{1}{2}bh$$
$$A(x) = \frac{1}{2}(x)(\sqrt{36 - x^2})$$
$$A(x) = \frac{x\sqrt{36 - x^2}}{2}$$

33. A: $[g \circ f]x = g(f(x)) = g(2x + 4) = (2x + 4)^2 - 3(2x + 4) + 2 = 4x^2 + 16x + 16 - 6x - 12 + 2 = 4x^2 + 10x + 6.$

34. C: One way to approach the problem is to use the table of values to first write equations for $f(x)$ and $g(x)$: $f(x) = 2x^2$ and $g(x) = 2x + 5$. Then, use those equations to find $f(g(-4))$.
$$g(-4) = 2(-4) + 5 = -3$$
$$f(-3) = 2(-3)^2 = 18$$
So, $f(g(-4)) = 18.$

35. D: By definition, when $f(x)$ and $g(x)$ are inverse functions, $f(g(x)) = g(f(x)) = x$. So, $f(g(4)) = 4.$

36. B: To find the inverse of an equation, solve for x in terms of y; then, exchange the variables x and y. Or, to determine if two functions $f(x)$ and $g(x)$ are inverses, find $f(g(x))$ and $g(f(x))$; if both results are x, then $f(x)$ and $g(x)$ are inverse functions.

For example, to find the inverse of $y = x + 6$, rewrite the equation $x = y + 6$ and solve for y. Since $y = x - 6$, the two given equations given in Choice A are inverses. Likewise, to find the inverse of $y = \frac{2x+3}{x-1}$, rewrite the equation as $x = \frac{2y+3}{y-1}$ and solve for y:
$$xy - x = 2y + 3$$
$$xy - 2y = x + 3$$
$$y(x - 2) = x + 3$$
$$y = \frac{x + 3}{x - 2}$$

The two equations given in Choice C are inverses.

Here, the second method is used to determine if the two equations given in Choices B and D are inverses:
Choice B: $y = 2(2x + 3) - 3 = 4x + 6$. The two given equations are **NOT** inverses. Choice D: $y = \frac{(2x+1)-1}{2} = \frac{2x}{2} = x$ and $y = 2\left(\frac{x-1}{2}\right) + 1 = x - 1 + 1 = x$, so the two given equations are inverses.

37. A: Below is the graph of $g(x)$.

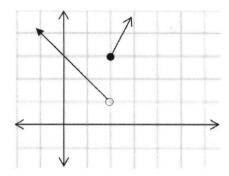

Statement II is true: the graph is indeed discontinuous at $x = 2$. Since $g(3) = 2(3) - 1 = 5$, Statement I is false, and since the range is $y > 1$, Statement III is also false.

38. A: In the range $(-\infty, -1)$, the graph represented is $y = x^2$. In the range $[-1, 2)$, the graph is the greatest integer function, $y = [\![x]\!]$. In the range $[-2, \infty)$, the graph is $y = -2x + 6$.

39. B: If $y = a(x + b)(x + c)^2$, the degree of the polynomial is 3. Since the degree of the polynomial is odd and the leading coefficient is positive $(a > 0)$, the end behavior of the graph is below.

$$\swarrow \nearrow$$

Therefore, neither Choice A nor Choice C can be a graph of $y = a(x + b)(x + c)^2$. The maximum number of "bumps" (or critical points) in the graph is at most one less than the degree of the polynomial, so Choice D, which has three bumps, cannot be the graph of the function. Choice B displays the correct end behavior and has two bumps, so it is a possible graph of $y = a(x + b)(x + c)^2$.

40. B: $5n + 3s \geq 300$ when $n =$ number of non-student tickets which must be sold and $s =$ number of student tickets which must be sold. The intercepts of this linear inequality are $n = 60$ and $s = 100$. The solid line through the two intercepts represents the minimum number of each type of ticket which must be sold in order to offset production costs. All points above the line represent sales which result in a profit for the school.

41. D: The vertex form of a quadratic equation is $y = a(x - h)^2 + k$, where $x = h$ is the parabola's axis of symmetry and (h, k) is the parabola's vertex. The vertex of the graph is (-1,3), so the equation can be written as $y = a(x + 1)^2 + 3$. The parabola passes through point (1,1), so $1 = a(1 + 1)^2 + 3$. Solve for a:
$$1 = a(1 + 1)^2 + 3$$
$$1 = a(2)^2 + 3$$
$$1 = 4a + 3$$
$$-2 = 4a$$
$$-\frac{1}{2} = a$$

So, the vertex form of the parabola is $y = -\frac{1}{2}(x + 1)^2 + 3$. Write the equation in the form $y = ax^2 + bx + c$.

$$y = -\frac{1}{2}(x + 1)^2 + 3$$
$$y = -\frac{1}{2}(x^2 + 2x + 1) + 3$$
$$y = -\frac{1}{2}x^2 - x - \frac{1}{2} + 3$$
$$y = -\frac{1}{2}x^2 - x + \frac{5}{2}$$

42. D: There are many ways to solve quadratic equations in the form $ax^2 + bx + c = 0$; however, some methods, such as graphing and factoring, may not be useful for some equations, such as those with irrational or complex roots. Solve this equation by completing the square or by using the Quadratic Formula, $x = \frac{-b \pm \sqrt{b^2 - 4ac}}{2a}$.

$$7x^2 + 6x + 2 = 0; a = 7, b = 6, c = 2$$

$$x = \frac{-b \pm \sqrt{b^2 - 4ac}}{2a}$$
$$x = \frac{-6 \pm \sqrt{6^2 - 4(7)(2)}}{2(7)}$$
$$x = \frac{-6 \pm \sqrt{36 - 56}}{14}$$
$$x = \frac{-6 \pm \sqrt{-20}}{14}$$
$$x = \frac{-6 \pm 2i\sqrt{5}}{14}$$
$$x = \frac{-3 \pm i\sqrt{5}}{7}$$

43. C: A system of linear equations can be solved by using matrices or by using the graphing, substitution, or elimination (also called linear combination) method. The elimination method is shown here:

$$3x + 4y = 2$$
$$2x + 6y = -2$$

In order to eliminate x by linear combination, multiply the top equation by 2 and the bottom equation by –3 so that the coefficients of the x-terms will be additive inverses.

$$2(3x + 4y = 2)$$
$$-3(2x + 6y = -2)$$

Then, add the two equations and solve for y.

$$6x + 8y = 4$$
$$\underline{-6x - 18y = 6}$$
$$-10y = 10$$
$$y = -1$$

Substitute -1 for y in either of the given equations and solve for x.

$$3x + 4(-1) = 2$$
$$3x - 4 = 2$$
$$3x = 6$$
$$x = 2$$

The solution to the system of equations is $(2, -1)$.

44. C: The graph below shows that the lines are parallel and that the shaded regions do not overlap. There is no solution to the set of inequalities given in Choice C.

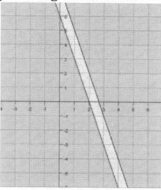

$6x + 2y \leq 12$
$2y \leq -6x + 12$
$y \leq -3x + 6$

$3x \geq 8 - y$
$y \geq -3x + 8$

As in Choice C, the two lines given in Choice A are parallel; however, the shading overlaps between the lines, so that region represents the solution to the system of inequalities.

The shaded regions for the two lines in Choice B do not overlap except at the boundary, but since the boundary is same, the solution to the system of inequalities is the line $y = -2x + 6$.

Choice D contains a set of inequalities which have intersecting shaded regions; the intersection represents the solution to the system of inequalities.

45. A: First, write three equations from the information given in the problem. Since the total number of tickets sold was 810, $x + y + z = 810$. The ticket sales generated \$14,500, so $15x + 25y + 20z = 14,500$. The number of children under ten was the same as twice the number of adults and seniors, so $x = 2(y + z)$, which can be rewritten as $x - 2y - 2z = 0$.

The coefficients of each equation are arranged in the rows of a 3x3 matrix, which, when multiplied by the 3x1 matrix arranging the variables x, y, and z, will give the 3x1 matrix which arranges the constants of the equations.

46. B: There are many ways to solve this system of equations. One is shown below.
1. Multiply the second equation by 2 and combine it with the first equation to eliminate the variable y.

$$2x - 4y + z = 10$$
$$\underline{-6x + 4y - 8z = -14}$$
$$-4x - 7z = -4$$

2. Multiply the third equation by –2 and combine it with the original second equation to eliminate y.

$$-3x + 2y - 4z = -7$$
$$\underline{-2x - 2y + 6z = 2}$$
$$-5x + 2z = -5$$

3. Multiply the equation from step one by 5 and the equation from step two by -4 and combine to eliminate x.

$$-20x - 35z = -20$$
$$\underline{20x - 8z = 20}$$
$$-43z = 0$$
$$z = 0$$

4. Substitute 0 for z in the equation from step 2 to find x.

$$-5x + 2(0) = -5$$
$$-5x = -5$$
$$x = 1$$

5. Substitute 0 for z and 1 for x into the first original equation to find y.

$$2(1) - 4y + (0) = 10$$
$$2 - 4y = 10$$
$$-4y = 8$$
$$y = -2$$

47. B: One way to solve the equation is to write $x^4 + 64 = 20x^2$ in the quadratic form $(x^2)^2 - 20(x^2) + 64 = 0$. This trinomial can be factored as $(x^2 - 4)(x^2 - 16) = 0$. In each set of parentheses is a difference of squares, which can be factored further: $(x + 2)(x - 2)(x + 4)(x - 4) = 0$. Use the zero product propery to find the solutions to the equation.

$$x + 2 = 0 \qquad x - 2 = 0 \qquad x + 4 = 0 \qquad x - 4 = 0$$
$$x = -2 \qquad\quad x = 2 \qquad\quad x = -4 \qquad\quad x = 4$$

48. A: First, set the equation equal to zero.

$$3x^3y^2 - 45x^2y = 15x^3y - 9x^2y^2$$
$$3x^3y^2 - 15x^3y + 9x^2y^2 - 45x^2y = 0$$

Then, factor the equation.

$$3x^2y(xy - 5x + 3y - 15) = 0$$
$$3x^2y[x(y - 5) + 3(y - 5)] = 0$$
$$3x^2y[(y - 5)(x + 3)] = 0$$

Use the zero product property to find the solutions.

$$3x^2y = 0 \qquad y - 5 = 0 \qquad x + 3 = 0$$
$$x = 0 \qquad\quad y = 5 \qquad\quad x = -3$$
$$y = 0$$

So, the solutions are $x = \{0, -3\}$ and $y = \{0, 5\}$.

49. D: The degree of $f(x)$ is 1, the degree of $g(x)$ is 2, and the degree of $h(x)$ is 3. The leading coefficient for each function is 2. Functions $f(x)$ and $h(x)$ have exactly one real zero ($x = 1$), while $g(x)$ has two real zeros ($x = \pm 1$):

$f(x)$	$g(x)$	$h(x)$
$0 = 2x - 2$	$0 = 2x^2 - 2$	$0 = 2x^3 - 2$
$-2x = -2$	$-2x^2 = -2$	$-2x^3 = -2$
$x = 1$	$x^2 = 1$	$x^3 = 1$
	$x = 1;\ x = -1$	$x = 1$

50. B: The path of a bullet is a parabola, which is the graph of a quadratic function. The path of a sound wave can be modeled by a sine or cosine function. The distance an object travels over time given a constant rate is a linear relationship, while radioactive decay is modeled by an exponential function.

51. B: First, use the properties of logarithms to rewrite $2 \log_4 y + \log_4 16 = 3$.
- Since $N \log_a M = \log_a M^N$, $2 \log_4 y = \log_4 y^2$. Replacing $2 \log_4 y$ by its equivalent in the given equation gives $\log_4 y^2 + \log_4 16 = 3$.
- Since $\log_a M + \log_a N = \log_a MN$, $\log_4 y^2 + \log_4 16 = \log_4 16\, y^2$. Thus, $\log_4 16\, y^2 = 3$.
- Since $\log_a M = N$ is equivalent to $a^N = M$, $\log_4 16\, y^2 = 3$ is equivalent to $4^3 = 16y^2$.

Then, solve for y. (Note that y must be greater than zero.)
$$4^3 = 16y^2$$
$$64 = 16y^2$$
$$4 = y^2$$
$$2 = y$$
Finally, substitute 2 for y in the expression $\log_y 256$ and simplify: $\log_2 256 = 8$ since $2^8 = 256$.

52. B: First, apply the laws of exponents to simplify the expression on the left.
$$\frac{(x^2y)(2xy^{-2})^3}{16x^5y^2} + \frac{3}{xy}$$

$$\frac{(x^2y)(8x^3y^{-6})}{16x^5y^2} + \frac{3}{xy}$$

$$\frac{8x^5y^{-5}}{16x^5y^2} + \frac{3}{xy}$$

$$\frac{1}{2y^7} + \frac{3}{xy}$$

Then, add the two fractions.

$$\frac{1}{2y^7} \cdot \frac{x}{x} + \frac{3}{xy} \cdot \frac{2y^6}{2y^6}$$

$$\frac{x}{2xy^7} + \frac{6y^6}{2xy^7}$$

$$\frac{x + 6y^6}{2xy^7}$$

53. C: If $f(x) = 10^x$ and $f(x) = 5$, then $5 = 10^x$. Since $\log_{10}x$ is the inverse of 10^x, $\log_{10}5 = \log_{10}(10^x) = x$. Therefore, $0.7 \approx x$.

54. C: The graph shown is the exponential function $y = 2^x$. Notice that the graph passes through (-2, 0.25), (0,1), (2,4).

x	Choice A x^2	Choice B $\sqrt{x}$	Choice C 2^x	Choice D $\log_2 x$
-2	4	undefined in $\mathbb{R}$	0.25	undefined
0	0	0	1	undefined
2	4	$\sqrt{2}$	4	1

55. C: The x-intercept is the point at which $f(x) = 0$. When $0 = \log_b x$, $b^0 = x$; since $b^0 = 1$, the x-intercept of $f(x) = \log_b x$ is always 1. If $f(x) = \log_b x$ and $x = b$, then $f(x) = \log_b b$, which is, by definition, 1. ($b^1 = b$.) If $g(x) = b^x$, then $f(x)$ and g(x) are inverse functions and are therefore symmetric with respect to $y = x$. The statement choice C is not necessarily true since $x < 1$ includes numbers less than or equal to zero, the values for which the function is undefined. The statement $f(x) < 0$ is true only for x values between 0 and 1 ($0 < x < 1$).

56. D: Bacterial growth is exponential. Let x be the number of doubling times and a be the number of bacteria in the colony originally transferred into the broth and y be the number of bacteria in the broth after a doubling times.

Time	Number of doubling times (x)	$a(2^x)$	Number of bacteria (y)
0	0	$a(2^0) = a$	1×10^6
20 minutes	1	$a(2^1)$	2×10^6
40 minutes	2	$a(2^2)$	4×10^6
60 minutes	3	$\boldsymbol{a(2^3)}$	$\mathbf{8 \times 10^6}$

Determine how many bacteria were present in the original colony. Either work backwards by halving the number of bacteria (see gray arrows above) or calculate a:
$$a(2^3) = 8 \times 10^6$$
$$8a = 8 \times 10^6$$
$$a = 10^6$$
The equation for determining the number of bacteria is $y = (2^x) \cdot 10^6$. Since the bacteria double every twenty minutes, they go through three doubling times every hour. So, when the bacteria are allowed to grow for eight hours, they will have gone through 24 doubling times. When $x = 24$, $y = (2^{24}) \cdot 10^6 = 16777216 \times 10^6$, which is approximately 1.7×10^{13}.

57. C: Since the pH scale is a base–10 logarithmic scale, a difference in pH of 1 indicates a ratio between strengths of 10. So, an acid with a pH of 3 is 100 times stronger than an acid with a pH of 5.

58. A:

$$\sqrt{\frac{-28x^6}{27y^5}} = \frac{2x^3 i\sqrt{7}}{3y^2\sqrt{3y}} \cdot \frac{\sqrt{3y}}{\sqrt{3y}} = \frac{2x^3 i\sqrt{21y}}{9y^2}$$

59. C:

$-4 \leq 2 + 3(x-1) \leq 2$ $-6 \leq 3(x-1) \leq 0$ $-2 \leq x - 1 \leq 0$ $-1 \leq x \leq 1$	$-2x^2 + 2 \geq x^2 - 1$ $-3x^2 \geq -3$ $x^2 \leq 1$ $-1 \leq x \leq 1$	$\dfrac{11 - |3x|}{7} \geq 2$ $11 - |3x| \geq 14$ $-|3x| \geq 3$ $|3x| \leq -1$ No solution	$3|2x| + 4 \leq 10$ $3|2x| \leq 6$ $|2x| \leq 2$ $-2 \leq 2x \leq 2$ $-1 \leq x \leq 1$

60. D: When solving radical equations, check for extraneous solutions.

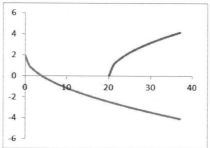

$2 - \sqrt{x} = \sqrt{x - 20}$
$\left(2 - \sqrt{x}\right)^2 = \left(\sqrt{x - 20}\right)^2$
$4 - 4\sqrt{x} + x = x - 20$
$-4\sqrt{x} = -24$
$\sqrt{x} = 6$
$\sqrt{x}^2 = 6^2$
$x = 36$

$2 - \sqrt{36} = \sqrt{36 - 20}$
$2 - 6 = \sqrt{16}$
$-4 \neq 4$

Since the solution does not check, there is no solution. Notice that the graphs $y = 2 - \sqrt{x}$ and $y = \sqrt{x - 20}$ do not intersect, which confirms there is no solution.

61. B: Notice that choice C cannot be correct since $x \neq 1$. ($x = 1$ results in a zero in the denominator.)

$$\frac{x-2}{x-1} = \frac{x-1}{x+1} + \frac{2}{x-1}$$
$$(x-1)(x+1)\left(\frac{x-2}{x-1} = \frac{x-1}{x+1} + \frac{2}{x-1}\right)$$
$$(x+1)(x-2) = (x-1)^2 + 2(x+1)$$
$$x^2 - x - 2 = x^2 - 2x + 1 + 2x + 2$$
$$x^2 - x - 2 = x^2 + 3$$
$$-x = 5$$
$$x = -5$$

62. A: The denominator of a fraction cannot equal zero. Therefore, for choices A and B,.
$$x^2 - x - 2 \neq 0$$
$$(x+1)(x-2) \neq 0$$
$$x + 1 \neq 0 \quad x - 2 \neq 0$$
$$x \neq -1 \quad x \neq 2.$$

Since choice A is in its simplest form, there are vertical asymptotes at $x = -1$ and $x = 2$. However, for choice B,
$$\frac{3x+3}{x^2 - x - 2} = \frac{3(x+1)}{(x+1)(x-2)} = \frac{3}{x-2}.$$
So, at $x = -2$ there is an asymptote, while at $x = -1$, there is simply a hole in the graph. So, choice B does not match the given graph. For choice C, there are asymptotes at $x = -1$ and $x = 2$; however, notice that it is possible for the graph of choice C to intersect the x-axis since it is possible that $y = 0$ (when $x = 0.5$). Since the given graph does not have an x-intercept, choice C is incorrect. For choice A, it is not possible that y=0, so it is a possible answer. Check a few points on the graph to make sure they satisfy the equation.

x	y
-2	$\frac{3}{4}$
0	$-\frac{3}{2}$
$\frac{1}{2}$	$-\frac{4}{3}$
1	$-\frac{3}{2}$
3	$\frac{3}{4}$

The points $\left(-2, \frac{3}{4}\right), \left(0, -\frac{3}{2}\right), \left(\frac{1}{2}, -\frac{4}{3}\right), \left(1, -\frac{3}{2}\right)$, and $(3, \frac{3}{4})$ are indeed points on the graph.

63. A: An easy way to determine which is the graph of $f(x) = -2|-x+4| - 1$ is to find $f(x)$ for a few values of x. For example, $f(x) = -2|0+4| - 1 = -9$. Graphs A and B pass through $(0, -9)$, but graphs C and D do not. $f(4) = -2|-4+4| - 1 = -1$. Graphs A and D pass through $(4, -1)$, but graphs B and C do not. Graph A is the correct graph. $f(x) = -2|-x+4| - 1$ shifts the graph of $y = |x|$ to the left four units, reflects it across the y-axis, inverts it, makes it narrower, and shifts it down one unit.

64. C: The first function shifts the graph of $y = \frac{1}{x}$ to the right one unit and up one unit. The domain and range of $y = \frac{1}{x}$ are $\{x: x \neq 0\}$ and $\{y: y \neq 0\}$, so the domain and range of $y = \frac{1}{x-1} + 1$ are $\{x: x \neq 1\}$ and $\{y: y \neq 1\}$. The element 1 is not in its domain.

The second function inverts the graph of $y = \sqrt{x}$ and shifts it to the left two units and down one unit. The domain and range of $y = \sqrt{x}$ are $\{x: x \geq 0\}$ and $\{y: y \geq 0\}$, so the domain and range of $y = -\sqrt{x + 2} - 1$ are $\{x: x \geq -2\}$ and $\{y: y \leq -1\}$. The range does not contain the element 2.

The third function shifts the graph of $y = |x|$ to the left two units and down three units. The domain of $y = |x|$ the set of all real numbers and range is $\{y: y \geq 0\}$, so the domain of $y = |x + 2| - 3$ is the set of all real numbers and the range is $\{y: y \geq -3\}$. The domain contains the element 1 and the range contains the element 2.

This is the graph of the fourth function. The domain of this piece-wise function is the set of all real numbers, and the range is $\{y: y \leq -1\}$. The range does not contain the element 2.

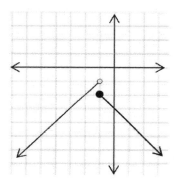

65. B: First, state the exclusions of the domain.
$$x^3 + 2x^2 - x - 2 \neq 0$$
$$(x + 2)(x - 1)(x + 1) \neq 0$$
$$x + 2 \neq 0 \quad x - 1 \neq 0 \quad x + 1 \neq 0$$
$$x \neq -2 \quad x \neq 1 \quad x \neq -1$$

To determine whether there are asymptotes or holes at these values of x, simplify the expression $\frac{x^2-x-6}{x^3+2x^2-x-2}$.
$$\frac{(x - 3)(x + 2)}{(x + 2)(x - 1)(x + 1)} = \frac{x - 3}{(x - 1)(x + 1)}$$
There are asymptotes at $x = 1$ and at $x = -1$ and a hole at $x = 2$. Statement I is false.

To find the x-intercept of $f(x)$, solve $f(x) = 0$. $f(x) = 0$ when the numerator is equal to zero. The numerator equals zero when $x = 2$ and $x = 3$; however, 2 is excluded from the domain of $f(x)$, so the x-intercept is 3. To find the y-intercept of $f(x)$, find $f(0)$. $\frac{0^2-0-6}{0^3+2(0)^2-0-2} = \frac{-6}{-2} = 3$. The y-intercept is 3. Statement II is true.

66. C: The period of the pendulum is a function of the square root of the length of its string, and is independent of the mass of the pendulum or the angle from which it is released. If the period of

Pendulum 1's swing is four times the period of Pendulum 2's swing, then the length of Pendulum 1's string must be 16 times the length of Pendulum 2's swing since all other values besides L in the expression $2\pi\sqrt{\dfrac{L}{g}}$ remain the same.

67. D: There are many ways Josephine may have applied her knowledge to determine how to approximately measure her medicine using her plastic spoon. The only choice which correctly uses dimensional analysis is choice D: the dosage ≈ 25 cc $\cdot\dfrac{1\text{ ml}}{1\text{ cc}}\cdot\dfrac{1\text{L}}{1000\text{ml}}\cdot\dfrac{0.5\text{ gal}}{2\text{L}}\cdot\dfrac{16\text{c}}{1\text{ gal}}\cdot\dfrac{48\text{t}}{1\text{c}}\cdot\dfrac{1\text{ spoonful}}{1\text{t}}$

$\rightarrow\dfrac{25}{1000}\cdot\dfrac{1}{4}\cdot 16\cdot 48\approx 5$.

68. C: If 1" represents 60 feet, 10" represents 600 ft, which is the same as 200 yards.

69. D: If the distance between the two houses is 10 cm on the map, then the actual distance between the houses is 100 m.

To find x, use the Pythagorean Theorem:

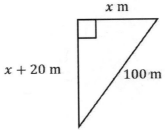

$x^2 + (x+20)^2 = (100)^2$
$x^2 + x^2 + 40x + 400 = 10000$
$2x^2 + 40x - 9600 = 0$
$2(x^2 + 20x - 4800) = 0$
$2(x-60)(x+80) = 0$
$x = 60 \quad x = -80$

Since x represents a distance, it cannot equal –80. Since $x = 60$, $x + 20 = 80$. Roxana walks a total of 140 m to get to her friend's house.

70. D: $\triangle ABC$ is similar to the smaller triangle with which it shares vertex A. $AB = (2x-1) + (x+7) = 3x+6$. $AC = 4+8 = 12$. Set up a proportion and solve for x:
$$\frac{3x+6}{12} = \frac{2x-1}{4}$$
$$12x + 24 = 24x - 12$$
$$36 = 12x$$
$$3 = x$$

So, $AB = 3x + 6 = 3(3) + 6 = 15$.

71. B: Percent error $= \frac{|\text{actual value}-\text{measured value}|}{\text{actual value}} \times 100\%$, and the average percent error is the sum of the percent errors for each trial divided by the number of trials.

	% error Trial 1	% error Trial 2	% error Trial 3	% error Trial 4	Average percent error
Scale 1	0.1%	0.2%	0.2%	0.1%	0.15%
Scale 2	2.06%	2.09%	2.10%	2.08%	2.08%

The percent error for Scale 1 is less than the percent error for Scale 2, so it is more accurate. The more precise scale is Scale 2 because its range of values, $10.210 \text{ g} - 10.206 \text{ g} = 0.004 \text{ g}$, is smaller than the Scale 2's range of values, $10.02 \text{ g} - 9.98 \text{ g} = 0.04 \text{ g}$.

72. C: If l and w represent the length and width of the enclosed area, its perimeter is equal to $2l + 2w$; since the fence is positioned x feet from the lot's edges on each side, the perimeter of the lot is $2(l + 2x) + 2(w + 2x)$. Since the amount of money saved by fencing the smaller are is \$432, and since the fencing material costs \$12 per linear foot, 36 fewer feet of material are used to fence around the playground than would have been used to fence around the lot. This can be expressed as the equation $2(l + 2x) + 2(w + 2x) - (2l + 2w) = 36$.

$$2(l + 2x) + 2(w + 2x) - (2l + 2w) = 36$$
$$2l + 4x + 2w + 4x - 2l - 2w = 36$$
$$8x = 36$$
$$x = 4.5 \text{ ft}$$

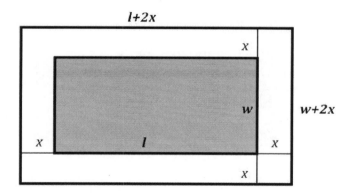

The difference in the area of the lot and the enclosed space is 141 yd^2, which is the same as 1269 ft^2. So, $(l + 2x)(w + 2x) - lw = 1269$. Substituting 4.5 for x,
$$(l + 9)(w + 9) - lw = 1269$$
$$lw + 9l + 9w + 81 - lw = 1269$$
$$9l + 9w = 1188$$
$$9(l + w) = 1188$$
$$l + w = 132 \text{ ft}$$

Therefore, the perimeter of the enclosed space, $2(l + w)$, is $2(132) = 264 \text{ ft}$. The cost of 264 ft of fencing is $264 \cdot \$12 = \$3{,}168$.

73. B: The volume of Natasha's tent is $\frac{x^2h}{3}$. If she were to increase by 1 ft the length of each side of the square base, the tent's volume would be $\frac{(x+1)^2h}{3} = \frac{(x^2+2x+1)(h)}{3} = \frac{x^2h+2xh+h}{3} = \frac{x^2h}{3} + \frac{2xh+h}{3}$. Notice this is the volume of Natasha's tent, $\frac{x^2h}{3}$, increased by $\frac{2xh+h}{3}$, or $\frac{h(2x+1)}{3}$.

74. A: The area of a circle is πr^2, so the area of a semicircle is $\frac{\pi r^2}{2}$. Illustrated below is a method which can be used to find the area of the shaded region.

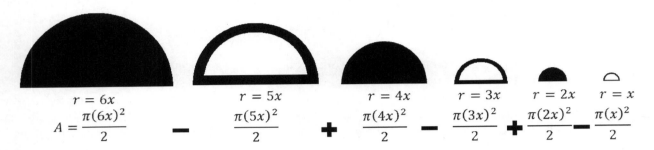

$$r = 6x \qquad r = 5x \qquad r = 4x \qquad r = 3x \qquad r = 2x \quad r = x$$

$$A = \frac{\pi(6x)^2}{2} \quad - \quad \frac{\pi(5x)^2}{2} \quad + \quad \frac{\pi(4x)^2}{2} \quad - \quad \frac{\pi(3x)^2}{2} \quad + \quad \frac{\pi(2x)^2}{2} \quad - \quad \frac{\pi(x)^2}{2}$$

The area of the shaded region is $\frac{\pi(36x^2-25x^2+16x^2-9x^2+4x^3-x^2)}{2} = \frac{(21x^2)\pi}{2}$.

75. B. Euclidean geometry is based on the flat plane. One of Euclid's five axioms, from which all Euclidean geometric theorems are derived, is the parallel postulate, which states that in a plane, for any line l and point A not on l, exactly one line which passes through A does not intersect l.

Non-Euclidean geometry considers lines on surfaces which are not flat. For instance, on the Earth's surface, if point A represents the North Pole and line l represents the equator (which does not pass through A), all lines of longitude pass through point A and intersect line l. In elliptical geometry, there are infinitely many lines which pass though A and intersect l, and there is no line which passes through A which does not also intersect l. In hyperbolic geometry, the opposite is true. When A is not on l, all lines which pass through A diverge from l, so none of the lines through A intersect l.

76. B: When four congruent isosceles trapezoids are arranged in an arch, the bases of the trapezoid come together to form regular octagons, the smaller of which is shown to the right. The measure of each angle of a regular octagon is 135°. $\left(\frac{(8-2)(180°)}{8} = 135°. \right)$ From the relationship of two of the trapezoid's base angles with one of the octagon's interior angles, write and solve an equation:

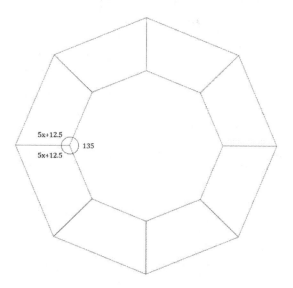

$(5x + 12.5) + (5x + 12.5) + 135 = 360$
$10x + 160 = 360$
$10x = 200$
$x = 20$

77. C: Sketch a diagram (this one is not to scale) and label the known segments. Use the property that two segments are congruent when they originate from the same point outside of a circle and are tangent to the circle.

The point of tangency of $\overline{CB}$ divides the segment into two pieces measuring 4 and 6; the point of tangency of $\overline{BA}$ divides the segment into two pieces measuring 6 and 8; the point of tangency of $\overline{AD}$ divides the segment into two pieces measuring 8 and 4. Therefore $AD = 8 + 4 = 12$.

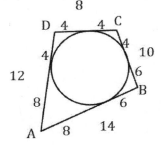

78. D: Let b represent the base of the triangle. The height h of the triangle is the altitude drawn from the vertex opposite of b to side b.

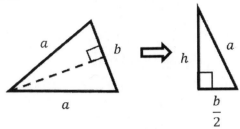

The height of the triangle can be found in terms of a and b by using the Pythagorean theorem:

$$h^2 + \left(\frac{b}{2}\right)^2 = a^2$$

$$h = \sqrt{a^2 - \frac{b^2}{4}} = \sqrt{\frac{4a^2 - b^2}{4}} = \frac{\sqrt{4a^2 - b^2}}{2}$$

The area of a triangle is $A = \frac{1}{2}bh$, so $A = \frac{1}{2}b\left(\frac{\sqrt{4a^2-b^2}}{2}\right) = \frac{b\sqrt{4a^2-b^2}}{4}$.

79. B: Since $\angle ADC$ is a right triangle with legs measuring 5 and 12, its hypotenuse measures 13. (5-12-13 is a Pythagorean triple.)

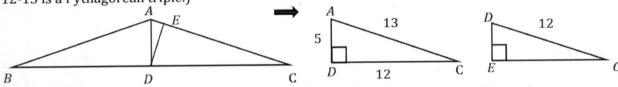

$\angle ADC$ and $\angle DEC$ are both right triangles which share vertex C. By the AA similarity theorem $\angle ADC \sim \angle DEC$. Therefore, a proportion can be written and solved to find DE.

$$\frac{5}{DE} = \frac{13}{12}$$
$$DE = 4.6$$

80. C: The center of the sphere is shared by the center of the cube, and each of the corners of the cube touches the surface of the sphere. Therefore, the diameter of the sphere is the line which passes through the center of the cube and connects one corner of the cube to the opposite corner on the opposite face. Notice in the illustration below that the diameter d of the sphere can be represented as the hypotenuse of a right triangle with a short leg measuring 4 units. (Since the volume of the cube is 64 cubic units, each of its sides measures $\sqrt[3]{64} = 4$ units.) The long leg of the triangle is the diagonal of the base of the cube. Its length can be found using the Pythagorean theorem: $4^2 + 4^2 = x^2$; $x = \sqrt{32} = 4\sqrt{2}$.

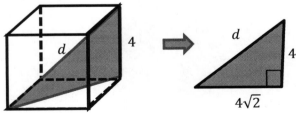

Use the Pythagorean theorem again to find d, the diameter of the sphere: $d^2 = \left(4\sqrt{2}\right)^2 + 4^2$; $d = \sqrt{48} = 4\sqrt{3}$. To find the volume of the sphere, use the formula $V = \frac{4}{3}\pi r^3$. Since the radius r of the sphere is half the diameter, $r = 2\sqrt{3}$, and $V = \frac{4}{3}\pi(2\sqrt{3})^3 = \frac{4}{3}\pi(24\sqrt{3}) = 32\pi\sqrt{3}$ cubic units.

81. D. Since it is given that $\overline{FD} \cong \overline{BC}$ and $\overline{AB} \cong \overline{DE}$, step 2 needs to establish either that $\overline{AC} \cong \overline{EF}$ or that $\triangle ABC \cong \triangle FDE$ in order for step 5 to show that $\triangle ABC \cong \triangle EDF$. The statement $\overline{AC} \cong \overline{EF}$ cannot be shown directly from the given information. On the other hand, $\triangle ABC \cong \triangle FDE$ can be determined: when two parallel lines ($\overline{BC}\|\overline{FG}$) are cut by a transversal ($\overline{AE}$), alternate exterior angles ($\triangle ABC$, $\triangle FDE$) are congruent. Therefore, $\triangle ABC \cong \triangle EDF$ by the side-angle-side (SAS) theorem.

82. A: Step 5 established that $\triangle ABC \cong \triangle EDF$. Because corresponding parts of congruent triangles are congruent (CPCTC), $\angle BAC \cong \angle DEF$. This is useful to establish when trying to prove $\overline{FE}\|\overline{AG}$: when two lines ($\overline{FE}$ and $\overline{AG}$) are cut by a transversal ($\overline{AE}$) and alternate interior angles ($\angle BAC$, $\angle DEF$) are congruent, then the lines are parallel. The completed proof is shown immediately following.

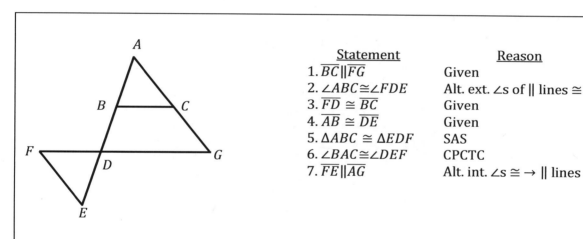

Statement	Reason
1. $\overline{BC}\|\overline{FG}$	Given
2. $\angle ABC \cong \angle FDE$	Alt. ext. $\angle$s of $\|$ lines $\cong$
3. $\overline{FD} \cong \overline{BC}$	Given
4. $\overline{AB} \cong \overline{DE}$	Given
5. $\triangle ABC \cong \triangle EDF$	SAS
6. $\angle BAC \cong \angle DEF$	CPCTC
7. $\overline{FE}\|\overline{AG}$	Alt. int. $\angle$s $\cong \rightarrow \|$ lines

Given: $\overline{BC}\|\overline{FG}$; $\overline{FD} \cong \overline{BC}$; $\overline{AB} \cong\overline{DE}$
Prove: $\overline{FE}\|\overline{AG}$

83. B: A cube has six square faces. The arrangement of these faces in a two-dimensional figure is a net of a cube if the figure can be folded to form a cube. Figures A, C, and D represent three of the eleven possible nets of a cube. If choice B is folded, however, the bottom square in the second column will overlap the fourth square in the top row, so the figure does not represent a net of a cube.

84. D: The cross-section is a hexagon.

85. A: Use the formula for the volume of a sphere to find the radius of the sphere:

$$V = \frac{4}{3}\pi r^3$$
$$36\pi = \frac{4}{3}\pi r^3$$
$$36 = \frac{4}{3}r^3$$
$$36 = \frac{4}{3}r^3$$
$$27 = r^3$$
$$3 = r$$

Then, substitute the point $(h, k, l) = (1, 0, -2)$ and the radius $r = 3$ into the equation of a sphere:

$$(x - h)^2 + (y - k)^2 + (z - l)^2 = r^2$$
$$(x - 1)^2 + y^2 + (z + 2)^2 = 3^2$$
$$(x - 1)^2 + y^2 + (z + 2)^2 = 9$$
$$x^2 - 2x + 1 + y^2 + z^2 + 4z + 4 = 9$$
$$x^2 + y^2 + z^2 - 2x + 4z = 4$$

86. B: The triangle is a right triangle with legs 3 and 4 units long.

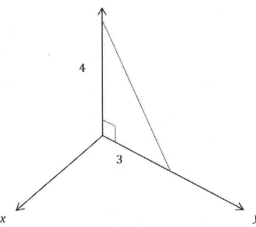

If the triangle is rotated about the z-axis, the solid formed is a cone with a height of 4 and a radius of 3; this cone has volume $V = \frac{1}{3}\pi r^2 h = \frac{1}{3}\pi 3^2 4 = 12\pi$ cubic units. If the triangle is rotated about the y-axis, the solid formed is a cone with a height of 3 and a radius of 4. This cone has volume

- 156 -

$V = \frac{1}{3}\pi r^2 h = \frac{1}{3}\pi 4^2 3 = 16\pi$ cubic units. The difference in the volumes of the two cones is $16\pi - 12\pi = 4\pi$ cubic units.

87. D: The point $(5, -5)$ lies on the line which has a slope of -2 and which passes through $(3, -1)$. If $(5, -5)$ is one of the endpoints of the line, the other would be $(1,3)$.

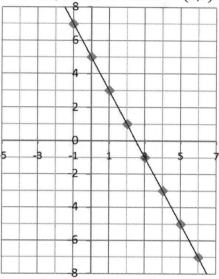

88. D: Since all of the answer choices are parallelograms, determine whether the parallelogram is also a rhombus or a rectangle or both. One way to do this is by examining the parallelogram's diagonals. If the parallelogram's diagonals are perpendicular, then the parallelogram is a rhombus. If the parallelogram's diagonals are congruent, then the parallelogram is a rectangle. If a parallelogram is both a rhombus and a rectangle, then it is a square.

To determine whether the diagonals are perpendicular, find the slopes of the diagonals of the quadrilateral:

- Diagonal 1: $\frac{6-2}{-5-3} = \frac{4}{-8} = -\frac{1}{2}$
- Diagonal 2: $\frac{0-8}{-3-1} = -\frac{8}{-4} = 2$

The diagonals have opposite inverse slopes and are therefore perpendicular. Thus, the parallelogram is a rhombus.

To determine whether the diagonals are congruent, find the lengths of the diagonals of the quadrilateral:

- Diagonal 1: $\sqrt{(6-2)^2 + (-5-3)^2} = \sqrt{(4)^2 + (-8)^2} = \sqrt{16 + 64} = \sqrt{80} = 4\sqrt{5}$
- Diagonal 2: $\sqrt{(0-8)^2 + (-3-1)^2} = \sqrt{(-8)^2 + (-4)^2} = \sqrt{64 + 16} = \sqrt{80} = 4\sqrt{5}$

The diagonals are congruent, so the parallelogram is a rectangle.

Since the polygon is a rhombus and a rectangle, it is also a square.

89. A: The equation of the circle is given in general form. When the equation is written in the standard form $(x - h)^2 + (y - k)^2 = r^2$, where (h, k) is the center of the circle and r is the radius of the circle, the radius is easy to determine. Putting the equation into standard form requires completing the square for x and y:

$$x^2 - 10x + y^2 + 8y = -29$$
$$(x^2 - 10x + 25) + (y^2 + 8y + 16) = -29 + 25 + 16$$
$$(x - 5)^2 + (y + 4)^2 = 12$$

Since $r^2 = 12$, and since r must be a positive number, $r = \sqrt{12} = 2\sqrt{3}$.

90. D: One way to determine whether the equation represents an ellipse, a circle, a parabola, or a hyperbola is to find the determinant $b^2 - 4ac$ of the general equation form of a conic section, $ax^2 + bxy + cy^2 + dx + ey + f = 0$, where $a, b, c, d, e,$ and f are constants. Given that the conic section is non-degenerate, if the determinant is positive, then the equation is a hyperbola; if the determinant is negative, then the equation is a circle (when $a = c$ and $b = 0$) or an ellipse; and if the determinant is zero, then the equation is a parabola. For $2x^2 - 3y^2 - 12x + 6y - 15 = 0$, $a = 2$, $b = 0$, $c = -3$, $d = -12$, $e = 6$, and $f = -15$. The determinant $b^2 - 4ac$ is equal to $0^2 - 4(2)(-3) = 24$. Since the determinant is positive, the graph is hyperbolic.

Another way to determine the shape of the graph is to look at the coefficients for the x^2 and y^2 terms in the given equation. If one of the coefficients is zero (in other words, if there is either an x^2 or a y^2 term in the equation but not both), then the equation is a parabola; if the coefficients have the same sign, then the graph is an ellipse or circle; and if the coefficients have opposite signs, then the graph is a hyperbola. Since the coefficient of x^2 is 2 and the coefficient of y^2 is -3, the graph is a hyperbola. That the equation can be written in the standard form for a hyperbola, $\frac{(x-h)^2}{a^2} - \frac{(y-k)^2}{b^2} = 1$, confirms the conclusion.

$$2x^2 - 3y^2 - 12x + 6y - 15 = 0$$
$$2x^2 - 12x - 3y^2 + 6y = 15$$
$$2(x^2 - 6x) - 3(y^2 - 2y) = 15$$
$$2(x^2 - 6x + 9) - 3(y^2 - 2y + 1) = 15 + 2(9) - 3(1)$$
$$2(x - 3)^2 - 3(y - 1)^2 = 30$$
$$\frac{(x - 3)^2}{15} - \frac{(y - 1)^2}{10} = 1$$

91. B: The graph of $f(x)$ is a parabola with a focus of (a, b) and a directrix of $y = -b$. The axis of symmetry of a parabola passes through the focus and vertex and is perpendicular to the directrix. Since the directrix is a horizontal line, the axis of symmetry is $x = a$; therefore, the x-coordinate of the parabola's vertex must be a. The distance between a point on the parabola and the directrix is equal to the distance between that point and the focus, so the y-coordinate of the vertex must be $y = \frac{-b+b}{2} = 0$. So, the vertex of the parabola given by $f(x)$ is $(a, 0)$.

If $g(x)$ were a translation of $f(x)$, as is the case for choices A, C, and D, the vertices of $f(x)$ and $g(x)$ would differ. Since the vertex of the graph of $g(x)$ is $(a, 0)$, none of those choices represent the correct response. However, if $g(x) = -f(x)$, the vertices of the graphs of both functions would be the same; therefore, this represents a possible relation between the two functions.

92. C: When a figure is reflected twice over non-parallel lines, the resulting transformation is a rotation about the point of intersection of the two lines of reflection. The two lines of reflection $y = x + 2$ and $x = 0$ intersect at $(0,2)$. So, $\Delta A''B''C''$ represents a rotation of ΔABC about the point $(0,2)$. The angle of rotation is equal to twice the angle between the two lines of reflection when measured in a clockwise direction from the first to the second line of reflection. Since the angle between the lines or reflection measures $135°$, the angle of rotation which is the composition of the two reflections measures $270°$. All of these properties can be visualized by drawing ΔABC, $\Delta A'B'C'$, and $\Delta A''B''C''$.

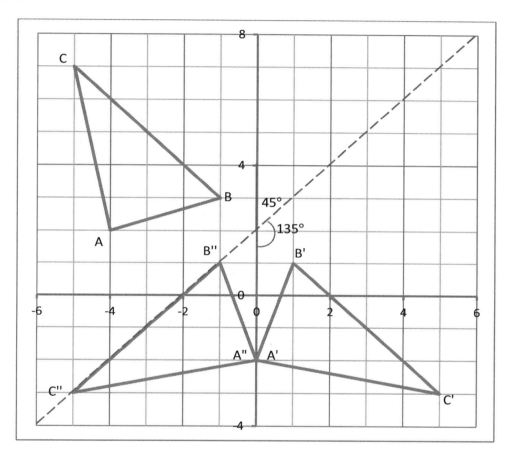

93. B: All regular polygons have rotational symmetry. The angle of rotation is the smallest angle by which the polygon can be rotated such that it maps onto itself; any multiple of this angle will also map the polygon onto itself. The angle of rotation for a regular polygon is the angle formed between two lines drawn from consecutives vertices to the center of the polygon. Since the vertices of a regular polygon lie on a circle, for a regular polygon with n sides, the angle of rotation measures $\frac{360°}{n}$.

Number of sides of regular polygon	Angle of rotation	Angles $\leq 360°$ which map the polygon onto itself
4	$\frac{360}{4} = 90°$	$90°, 180°, 270°, 360°$
6	$\frac{360}{6} = 60°$	$60°, 120°, 180°, 240°, 300°, 360°$
8	$\frac{360}{8} = 45°$	$45°, 90°, 135°, 180°, 225°, 270°, 315°, 316°$

- 159 -

10	$\dfrac{360}{10} = 36°$	$36°, 72°, 108°, 144°, 180°, 216°, 252°, 288°, 324°, 360°$

94. A: Since the y-coordinates of points P and Q are the same, line segment $\overline{PQ}$ is a horizontal line segment whose length is the difference in the x-coordinates a and c. Because the length of a line cannot be negative, and because it is unknown whether $a > c$ or $a < c$, $PQ = |a - c|$ or $|c - a|$. Since the x-coordinates of Q and Q' are the same, line segment $\overline{P'Q}$ is a vertical line segment whose length is $|d - b|$ or $|b - d|$. The quadrilateral formed by the transformation of $\overline{PQ}$ to $\overline{P'Q'}$ is a parallelogram. If the base of the parallelogram is $\overline{PQ}$, then the height is $\overline{P'Q}$ since $\overline{PQ} \perp \overline{P'Q}$. For a parallelogram, $A = bh$, so $A = |a - c| \cdot |b - d|$.

95. B: Since $\tan B = \dfrac{opposite}{adjacent} = \dfrac{b}{a}$, choice A is incorrect.

$\cos B = \dfrac{adjacent}{hypotenuse}$. The hypotenuse of a right triangle is equal to the square root of the sum of the squares of the legs, so $\cos B = \dfrac{adjacent}{hypotenuse} = \dfrac{a}{\sqrt{a^2+b^2}}$. Rationalize the denominator: $\dfrac{a}{\sqrt{a^2+b^2}} \cdot \dfrac{\sqrt{a^2+b^2}}{\sqrt{a^2+b^2}} = \dfrac{a\sqrt{a^2+b^2}}{a^2+b^2}$. Choice B is correct.

$\sec B = \dfrac{hypotenuse}{adjacent} = \dfrac{\sqrt{a^2+b^2}}{a}$, and $\csc B = \dfrac{\sqrt{a^2+b^2}}{b}$, so choices C and D are incorrect.

96. C: Find the missing angle measures in the diagram by using angle and triangle properties. Then, use the law of sines to find the distance y between the window and the wife's car: $\dfrac{60}{\sin 15°} = \dfrac{y}{\sin 45°}$, so $y = \dfrac{60 \sin 45°}{\sin 15} \approx 163.9$ ft. Use this number in a sine or cosine function to find x: $\sin 30° \approx \dfrac{x}{163.9}$, so $x \approx 163.9 \sin 30° \approx 82$. Therefore, the man's wife is parked approximately 82 feet from the building.

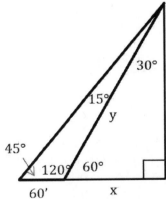

Alternatively, notice that when the man is looking down at a 45 degree angle, the triangle that is formed is an isosceles triangle, meaning that the height of his office is the same as the distance from the office to his car, or x + 60 feet. With this knowledge, the problem can be modeled with a single equation:

$$\dfrac{x + 60}{x} = \tan 60° \quad or \quad x = \dfrac{60}{\tan 60° - 1}$$

- 160 -

97. A: The reference angle for $-\frac{2\pi}{3}$ is $2\pi - \frac{2\pi}{3} = \frac{4\pi}{3}$, so $\tan(-\frac{2\pi}{3}) = \tan(\frac{4\pi}{3}) = \frac{\sin(\frac{4\pi}{3})}{\cos(\frac{4\pi}{3})}$. From the unit circle, the values of $\sin(\frac{4\pi}{3})$ and $\cos(\frac{4\pi}{3})$ are $-\frac{\sqrt{3}}{2}$ and $-\frac{1}{2}$, respectively. Therefore, $\tan(-\frac{2\pi}{3}) = \frac{-\frac{\sqrt{3}}{2}}{-\frac{1}{2}} = \sqrt{3}$.

98. D: On the unit circle, $\sin\theta = \frac{1}{2}$ when $\theta = \frac{\pi}{6}$ and when $\theta = \frac{5\pi}{6}$. Since only $\frac{5\pi}{6}$ is in the given range of $\frac{\pi}{2} < \theta < \pi$, $\theta = \frac{5\pi}{6}$.

99. C: Use trigonometric equalities and identities to simplify. $\cos\theta \cot\theta = \cos\theta \cdot \frac{\cos\theta}{\sin\theta} = \frac{\cos^2\theta}{\sin\theta} = \frac{1-\sin^2\theta}{\sin\theta} = \frac{1}{\sin\theta} - \sin\theta = \csc\theta - \sin\theta$.

100. B: The trigonometric identity $\sec^2\theta = \tan^2\theta + 1$ can be used to rewrite the equation $\sec^2\theta = 2\tan\theta$ as $\tan^2\theta + 1 = 2\tan\theta$, which can then be rearranged into the form $\tan^2\theta - 2\tan\theta + 1 = 0$. Solve by factoring and using the zero product property:
$$\tan^2\theta - 2\tan\theta + 1 = 0$$
$$(\tan\theta - 1)^2 = 0$$
$$\tan\theta - 1 = 0$$
$$\tan\theta = 1.$$
Since $\tan\theta = 1$ when $\sin\theta = \cos\theta$, for $0 < \theta \leq 2\pi$, $\theta = \frac{\pi}{4}$ or $\frac{5\pi}{4}$.

101. A: Since the graph shows a maximum height of 28 inches above the ground, and since the maximum distance from the road the pebble reaches is when it is at the top of the tire, the diameter of the tire is 28 inches. Therefore, its radius is 14 inches. From the graph, it can be observed that the tire makes 7.5 rotations in 0.5 seconds. Thus, the tire rotates 15 times in 1 second, or $15 \cdot 60 = 900$ times per minute.

102. C: The dashed line represents the sine function (x), and the solid line represents a cosine function $g(x)$. The amplitude of $f(x)$ is 4, and the amplitude of $g(x)$ is 2. The function $y = \sin x$ has a period of 2π, while the graph of function $f(x) = a_1 \sin(b_1 x)$ has a period of 4π; therefore, $b_1 = \frac{2\pi}{4\pi} = 0.5$, which is between 0 and 1. The graph of $g(x) = a_2 \cos(b_2 x)$ has a period of π, so $b_2 = \frac{2\pi}{\pi} = 2$.

103. B: The graph of $f(x)$ is stretched vertically by a factor of 4 with respect to $y = \sin x$, so $a_1 = 4$. The graph of $g(x)$ is stretched vertically by a factor of two and is inverted with respect to the graph of $y = \cos x$, so $a_2 = -2$. Therefore, the statement $a_2 < 0 < a_1$ is true.

104. A: The graph to the right shows the height h in inches of the weight on the spring above the table as a function of time t in seconds. Notice that the height is 3 in above the table at time 0 since the weight was pulled down two inches from its starting position 5 inches above the table. The spring fluctuates 2 inches above and below its equilibrium point, so its maximum height is 7 inches above the table. The graph represents a cosine curve which has been inverted, stretched vertically by a factor of 2, and shifted up five units; also, the graph has been compressed horizontally, with a period of 1 rather than 2π. So, the height of the weight on the spring as a function of time is $h = -2\cos(2\pi t) + 5$.

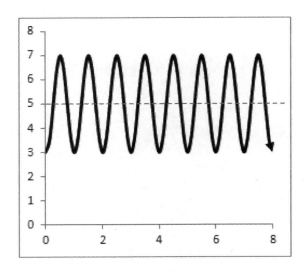

105. C: Since evaluating $\frac{x^3+3x^2-x-3}{x^2-9}$ at $x = -3$ produces a fraction with a zero denominator, simplify the polynomial expression before evaluating the limit:

$$\frac{x^3 + 3x^2 - x - 3}{x^2 - 9} = \frac{x^2(x+3) - 1(x+3)}{(x+3)(x-3)} = \frac{(x+3)(x^2-1)}{(x+3)(x-3)} = \frac{(x+1)(x-1)}{x-3}$$
$$\lim_{x \to -3} \frac{(x+1)(x-1)}{x-3} = \frac{(-3+1)(-3-1)}{-3-3} = \frac{8}{-6} = -\frac{4}{3}.$$

106. B: To evaluate the limit, divide the numerator and denominator by x^2 and use these properties of limits: $\lim_{x \to \infty} \frac{1}{x} = 0$; the limit of a sum of terms is the sum of the limits of the terms; and the limit of a product of terms is the product of the limits of the terms.

$$\lim_{x \to \infty} \frac{x^2 + 2x - 3}{2x^2 + 1} = \lim_{x \to \infty} \frac{\frac{x^2}{x^2} + \frac{2x}{x^2} - \frac{3}{x^2}}{\frac{2x^2}{x^2} + \frac{1}{x^2}} = \lim_{x \to \infty} \frac{1 + \frac{2}{x} - \frac{3}{x^2}}{2 + \frac{1}{x^2}} = \frac{1 + 0 - 0}{2 + 0} = \frac{1}{2}.$$

107. B: Evaluating $\frac{|x-3|}{3-x}$ when $x = 3$ produces a fraction with a zero denominator. To find the limit as x approaches 3 from the right, sketch a graph or make a table of values.

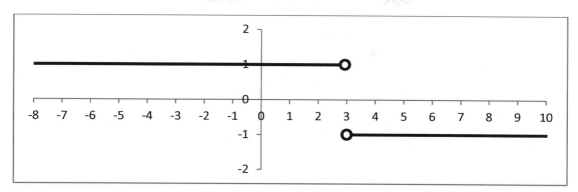

The value of the function approaches –1 as x approaches three from the right, so $\lim_{x \to 3^+} \frac{|x-3|}{3-x} = -1$.

108. C: The slope of the line tangent to the graph of a function f at $x = a$ is $f'(a)$. Since $f(x) = \frac{1}{4}x^2 - 3$, $f'(x) = 2\left(\frac{1}{4}\right)x^{(2-1)} - 0 = \frac{1}{2}x$. So, the slope at $x = 2$ is $f'(2) = \frac{1}{2}(2) = 1$.

109. D: The definition of the derivative of f at 2, or $f'(2)$, is the limit of the difference quotient $\lim_{h \to 0} \frac{f(2+h)-f(2)}{h}$. Rather than find the limit, simply evaluate the derivative of the function at $x = 2$:

$$f(x) = 2x^3 - 3x^2 + 4$$
$$f'(x) = 6x^2 - 6x$$
$$f'(2) = 6(2)^2 - 6(2)$$
$$f'(2) = 12$$

110. D: To find the derivative of $y = e^{3x^2-1}$, use the Chain Rule. Let $u = 3x^2 - 1$. Thus, $y = e^u$, and $\frac{dy}{du} = e^u$. Since $\frac{dy}{dx} = \frac{dy}{du} \cdot \frac{du}{dx}$, and since $\frac{du}{dx} = 6x$, $\frac{dy}{dx} = e^{3x^2-1} \cdot 6x = 6x\, e^{3x^2-1}$.

111. C: To find the derivative of $y = \ln(2x + 1)$, use the Chain Rule. Let $u = 2x + 1$. Thus, $y = \ln u$, and $\frac{dy}{du} = \frac{1}{u}$. Since $\frac{dy}{dx} = \frac{dy}{du} \cdot \frac{du}{dx}$, and since $\frac{du}{dx} = 2$, $\frac{dy}{dx} = \left(\frac{1}{2x+1}\right)(2) = \frac{2}{2x+1}$.

112. A: If $\lim_{x \to a^+} f(x) = \lim_{x \to a^-} f(x)$, then $\lim_{x \to a^+} f(x) = \lim_{x \to a^-} f(x) = \lim_{x \to a} f(x)$. Otherwise, $\lim_{x \to a} f(x)$ does not exist. If $\lim_{x \to a} f(x)$ exists, and if $\lim_{x \to a} f(x) = f(a)$, then the function is continuous at a. Otherwise, f is discontinuous at a.

113. A: To find the second derivative of the function, take the derivative of the first derivative of the function:

$$f(x) = 2x^4 - 4x^3 + 2x^2 - x + 1$$
$$f'(x) = 8x^3 - 12x^2 + 4x - 1$$
$$f''(x) = 24x^2 - 24x + 4.$$

114. A: The critical points of the graph occur when $f'(x) = 0$.
$$f(x) = 4x^3 - x^2 - 4x + 2$$
$$f'(x) = 12x^2 - 2x - 4$$
$$= 2(6x^2 - x - 2)$$
$$= 2(3x - 2)(2x + 1)$$

$$0 = 2(3x - 2)(2x + 1)$$
$$3x - 2 = 0 \quad 2x + 1 = 0$$
$$x = \frac{2}{3} \quad x = -\frac{1}{2}$$

If $f''(x) > 0$ for all x in an interval, the graph of the function is concave upward on that interval, and if $f''(x) < 0$ for all x in an interval, the graph of the function is concave upward on that interval. Find the second derivative of the function and determine the intervals in which $f''(x)$ is less than zero and greater than zero:
$$f''(x) = 24x - 2$$
$$24x - 2 < 0 \quad 24x - 2 > 0$$
$$x < \frac{1}{12} \quad x > \frac{1}{12}$$

The graph of f is concave downward on the interval $\left(-\infty, -\frac{1}{12}\right)$ and concave upward on the interval $\left(-\frac{1}{12}, \infty\right)$. The inflection point of the graph is $\left(\frac{1}{12}, f\left(\frac{1}{12}\right)\right) = \left(\frac{1}{12}, \frac{359}{216}\right)$. The point $\left(\frac{2}{3}, f\left(\frac{2}{3}\right)\right) = \left(\frac{2}{3}, \frac{2}{27}\right)$ is a relative minimum and the point $\left(-\frac{1}{2}, f\left(-\frac{1}{2}\right)\right) = \left(-\frac{1}{2}, 3\frac{1}{4}\right)$ is a relative maximum.

115. D: The velocity v of the ball at any time t is the slope of the line tangent to the graph of h at time t. The slope of a line tangent to the curve $h = -16t^2 + 50t + 3$ is h'.
$$h' = v = -32t + 50$$

When $t = 2$, the velocity of the ball is $-32(2) + 50 = -14$. The velocity is negative because the slope of the tangent line at $t = 2$ is negative; velocity has both magnitude and direction, so a velocity of -14 means that the velocity is 14 ft/s downward.

116. B: The manufacturer wishes to minimize the surface area A of the can while keeping its volume V fixed at 0.5 L = 500 mL = 500 cm^3. The formula for the surface area of a cylinder is $A = 2\pi rh + 2\pi r^2$, and the formula for volume is $V = \pi r^2 h$. To combine the two formulas into one, solve the volume formula for r or h and substitute the resulting expression into the surface area formula for r or h. The volume of the cylinder is 500 cm^3, so $500 = \pi r^2 h \rightarrow h = \frac{500}{\pi r^2}$. Therefore, $A = 2\pi rh + 2\pi r^2 \rightarrow 2\pi r \left(\frac{500}{\pi r^2}\right) + 2\pi r^2 = \frac{1000}{r} + 2\pi r^2$. Find the critical point(s) by setting the first derivative equal to zero and solving for r. Note that r represents the radius of the can and must therefore be a positive number.

$$A = 1000r^{-1} + 2\pi r^2$$
$$A' = -1000r^{-2} + 4\pi r$$
$$0 = -\frac{1000}{r^2} + 4\pi r$$
$$\frac{1000}{r^2} = 4\pi r$$
$$1000 = 4\pi r^3$$
$$\sqrt[3]{\frac{1000}{4\pi}} = r$$

So, when r≈4.3 cm, the minimum surface area is obtained. When the radius of the can is 4.30 cm, its height is $h \approx \frac{500}{\pi(4.30)^2} \approx 8.6$ cm, and the surface area is approximately $\frac{1000}{4.3} + 2\pi(4.3)^2 \approx 348.73$ cm^2. Confirm that the surface area is greater when the radius is slightly smaller or larger than 4.3 cm. For instance, when r=4 cm, the surface area is approximately 350.5 cm^2, and when r=4.5 cm, the surface area is approximately 349.5 cm^2.

117. C: Partitioned into rectangles with length of 1, the left Riemann sum is 20+25+28+30+29+26+22+16+12+10+10+13=241 square units, and the right Riemann sum is 25+28+30+29+26+22+16+12+10+10+13+17=238 square units.

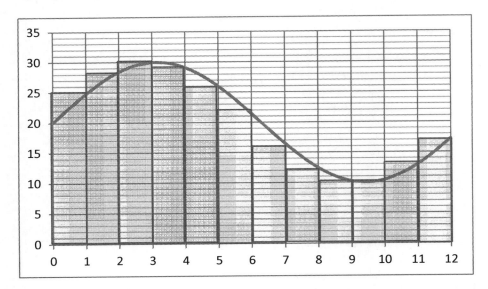

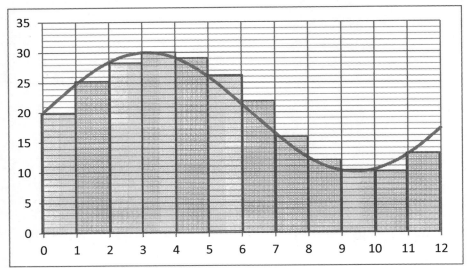

118. B: The area under curve $f(x)$ is $\int_1^2 \frac{1}{x} = [\ln(2)] - [\ln(1)] \approx 0.69$.

119. A: $\int 3x^2 + 2x - 1 = \frac{3}{2+1}x^{2+1} + \frac{2}{1+1}x^{1+1} - x + c = x^3 + x^2 - x + c$.

120. B: To calculate $\int 3x^2 e^{x^3} \, dx$, let $u = x^3$. Since $du = 3x^2 dx$, $\int 3x^2 e^{x^3} \, dx = \int e^u \, du \rightarrow e^u + c \rightarrow e^{x^3} + c$.

- 166 -

121. B: Find the points of intersection of the two graphs:
$$x^2 - 4 = -x + 2$$
$$x^2 + x - 6 = 0$$
$$(x + 3)(x - 2) = 0$$
$$x = -3 \quad x = 2$$

The finite region is bound at the top by the line $y = -x + 2$ and at the bottom by $y = x^2 - 4$, so the area is between the graphs on [-3,2], and the height of the region at point x is defined by $[(-x + 2) - (x^2 - 4)]$. Thus, the area of the region is

$$A = \int_{-3}^{2} [(-x + 2) - (x^2 - 4)]dx$$

$$= \int_{-3}^{2} (-x^2 - x + 6) \, dx$$

$$= \left[-\frac{1}{3}(2)^3 - \frac{1}{2}(2)^2 + 6(2)\right] - \left[-\frac{1}{3}(-3)^3 - \frac{1}{2}(-3)^2 + 6(-3)\right]$$

$$= \left[-\frac{8}{3} - 2 + 12\right] - \left[9 - \frac{9}{2} - 18\right] = \frac{22}{3} - \left(-\frac{27}{2}\right) = \frac{125}{6}$$

122. C: The acceleration a of an object at time t is the derivative of the velocity v of the object at time t, which is the derivative of the position x of the object at time t. So, given the velocity of an object at time t, $x(t)$ can be found by taking the integral of the $v(t)$, and $a(t)$ can be found by taking the derivative of $v(t)$.

$x(t) = \int v(t)dt = \int (12t - t^2)dt = 6t^2 - \frac{1}{3}t^3 + c$. Since the position of the car at time 0 is 0, $v(0) = 0 = 6(0)^2 - \frac{1}{3}(0)^3 + c \to 0 = 0 - 0 + c \to c = 0$. Therefore, $x(t) = 6t^2 - \frac{1}{3}t^3$.

$a(t) = v'(t) = 12 - 2t$.

Find the time at which the acceleration is equal to 0: $0 = 12 - 2t \to t = 6$. Then, find $x(6)$ to find the position of the car when the velocity is 0: $6(6)^2 - \frac{1}{3}(6)^3 = 216 - 72 = 144$.

123. D: To draw a box-and-whisker plot from the data, find the median, quartiles, and upper and lower limits.

```
3 | 6 7 9 9
4 | 2 3 8 8 9           Key
5 | 0 1 1 1 5 7       3|6 = 36
6 | 0 0 1 2 3
```

The median is $\frac{50+51}{2} = 50.5$, the lower quartile is $\frac{22+23}{2} = 22.5$, and the upper quartile is $\frac{57+60}{2} = 58.5$. The box of the box-and-whisker plot goes through the quartiles, and a line through the box represents the median of the data. The whiskers extend from the box to the lower and upper

- 167 -

limits, unless there are any outliers in the set. In this case, there are no outliers, so the box-and-whisker plot in choice A correctly represents the data set.

To draw a pie chart, find the percentage of data contained in each of the ranges shown. There are four out of twenty numbers between 30 and 39, inclusive, so the percentage shown in the pie chart for that range of data is $\frac{4}{20} \cdot 100\% = 20\%$; there are five values between 40 to 49, inclusive, so the percentage of data for that sector is $\frac{5}{20} \cdot 100\% = 25\%$; $\frac{6}{20} \cdot 100\% = 30\%$ of the data is within the range of 50-59, and $\frac{5}{20} \cdot 100\% = 25\%$ is within the range of 60-69. The pie chart shows the correct percentage of data in each category.

To draw a cumulative frequency histogram, find the cumulative frequency of the data.

Range	Frequency	Cumulative frequency
30-39	4	4
40-49	5	9
50-59	6	15
60-69	5	20

The histogram shows the correct cumulative frequencies.

Therefore, all of the graphs represent the data set.

124. B: A line graph is often used to show change over time. A Venn diagram shows the relationships among sets. A box and whisker plot shows displays how numeric data are distributed throughout the range. A pie chart shows the relationship of parts to a whole.

125. B: In choice A, the teacher surveys all the members of the population in which he is interested. However, since the response is voluntary, the survey is biased: the participants are self-selected rather than randomly selected. It may be that students who have a strong opinion are more likely to respond than those who are more neutral, and this would give the teacher a skewed perspective of student opinions. In choice B, students are randomly selected, so the sampling technique is not biased. In choice C, the student uses convenience sampling, which is a biased technique. For example, perhaps the student is in an honors class; his sampling method would not be representative of the entire class of eleventh graders, which includes both students who take and who do not take honors classes. Choice D also represents convenience sampling; only the opinions of parents in the PTA are examined, and these parents' opinions may not reflect the opinions of all parents of students at the school.

126. A: Nominal data are data that are collected which have no intrinsic quantity or order. For instance, a survey might ask the respondent to identify his or her gender. While it is possible to compare the relative frequency of each response (for example, "most of the respondents are women"), it is not possible to calculate the mean, which requires data to be numeric, or median, which requires data to be ordered. Interval data are both numeric and ordered, so mean and median can be determined, as can the mode, the interval within which there are the most data. Ordinal data has an inherent order, but there is not a set interval between two points. For example, a survey might ask whether the respondent whether he or she was very dissatisfied, dissatisfied, neutral, satisfied, or very satisfied with the customer service received. Since the data are not

numeric, the mean cannot be calculated, but since ordering the data is possible, the median has context.

127. A: The average number of male students in the 11th and 12th grades is 134 males. The number of Hispanic students at the school is 10% of 1219, which is 122 students. The difference in the number of male and female students at the school is $630 - 589 = 41$, and the difference in the number of 9th and 12th grade students at the school is $354 - 255 = 99$.

128. C: 52% of the student population is white. There are 630 female students at the school out of 1219 students, so the percentage of female students is $\frac{630}{1219} \cdot 100\% \approx 52\%$. The percentages rounded to the nearest whole number are the same.

129. D: 131 of 283 eleventh graders are male. Given that an 11th grader is chosen to attend the conference, the probability that a male is chosen is $\frac{\text{number of males}}{\text{number of 11th graders}} = \frac{131}{283} \approx 0.46$. Note that this is **NOT** the same question as one which asks for the probability of selecting at random from the school a male student who is in eleventh grade, which has a probability of $\frac{131}{1219} \approx 0.11$.

130. A: The range is the spread of the data. It can be calculated for each class by subtracting the lowest test score from the highest, or it can be determined visually from the graph. The difference between the highest and lowest test scores in class A is 98-23=75 points. The range for each of the other classes is much smaller.

131. D: 75% of the data in a set is above the first quartile. Since the first quartile for this set is 73, there is a 75% chance that a student chosen at random from class 2 scored above a 73.

132. C: The line through the center of the box represents the median. The median test score for classes 1 and 2 is 82.

Note that for class 1, the median is a better representation of the data than the mean. There are two outliers (points which lie outside of two standard deviations from the mean) which bring down the average test score. In cases such as this, the mean is not the best measure of central tendency.

133. D: Since there are 100 homes' market times represented in each set, the median time a home spends on the market is between the 50th and 51st data point in each set. The 50th and 51st data points for Zip Code 1 are six months and seven months, respectively, so the median time a house in Zip Code 1 spends on the market is between six and seven months (6.5 months), which by the realtor's definition of market time is a seven month market time. The 50th and 51st data points for Zip Code 2 are both thirteen months, so the median time a house in Zip Code 2 spends on the market is thirteen months.

To find the mean market time for 100 houses, find the sum of the market times and divide by 100. If the frequency of a one month market time is 9, the number 1 is added nine times (1·9), if frequency of a two month market time is 10, the number 2 is added ten times (2·10), and so on. So, to find the average market time, divide by 100 the sum of the products of each market time and its corresponding frequency. For Zip Code 1, the mean market time is 7.38 months, which by the realtor's definition of market time is an eight month market time. For Zip Code 2, the mean market time is 12.74, which by the realtor's definition of market time is a thirteen month market time.

The mode market time is the market time for which the frequency is the highest. For Zip Code 1, the mode market time is three months, and for Zip Code 2, the mode market time is eleven months.

The statement given in choice D is true. The median time a house spends on the market in Zip Code 1 is less than the mean time a house spends on the market in Zip Code 1.

Time on market	Frequency for Zip Code 1	Frequency for Zip Code 2	Time·Frequency for Zip Code 1	Time·Frequency for Zip Code 1
1	9	6	9	6
2	10	4	20	8
3	12	3	36	9
4	8	4	32	16
5	6	3	30	15
6	5	5	30	30
7	8	2	56	14
8	8	1	64	8
9	6	3	54	27
10	3	5	30	50
11	5	7	55	77
12	4	6	48	72
13	2	6	26	78
14	3	5	42	70
15	1	3	15	45
16	2	2	32	32
17	2	3	34	51
18	1	5	18	90
19	0	6	0	114
20	2	4	40	80
21	1	5	21	105
22	1	4	22	88
23	0	3	0	69
24	1	5	24	120
SUM	100	100	738	1274

134. C: The probability of an event is the number of possible occurrences of that event divided by the number of all possible outcomes. A camper who is at least eight years old can be eight, nine, or ten years old, so the probability of randomly selecting a camper at least eight years old is $\frac{\text{number of eight-, nine-, and ten-year old campers}}{\text{total number of campers}} = \frac{14+12+10}{12+15+14+12+10} = \frac{36}{63} = \frac{4}{7}$.

135. B: There are three ways in which two women from the same department can be selected: two women can be selected from the first department, or two women can be selected from the second department, or two women can be selected from the third department.

The probability that two women are selected from Department 1 is $\frac{12}{103} \times \frac{11}{102} = \frac{132}{10506}$, the probability that two women are selected from Department 2 is $\frac{28}{103} \times \frac{27}{102} = \frac{756}{10506}$, and the probability that two women are selected from Department 3 is $\frac{16}{103} \times \frac{15}{102} = \frac{240}{10506}$. Since any of these is a discrete possible outcome, the probability that two women will be selected from the same department is the sum of these outcomes: $\frac{132}{10506} + \frac{756}{10506} + \frac{240}{10506} \approx 0.107$, or 10.7%.

	Department 1	Department 2	Department 3	Total
Women	12	28	16	56
Men	18	14	15	47
Total	30	42	31	103

136. B: The number of students who like broccoli is equal to the number of students who like all three vegetables plus the number of students who like broccoli and carrots but not cauliflower plus the number of students who like broccoli and cauliflower but not carrots plus the number of students who like broccoli but no other vegetable: $3 + 15 + 4 + 10 = 32$. These students plus the numbers of students who like just cauliflower, just carrots, cauliflower and carrots, or none of the vegetables represents the entire set of students sampled: $32 + 2 + 27 + 6 + 23 = 90$. So, the probability that a randomly chosen student likes broccoli is $\frac{32}{90} \approx 0.356$.

The number of students who like carrots and at least one other vegetable is $15 + 6 + 3 = 24$. The number of students who like carrots is $24 + 27 = 51$. So, the probability that a student who likes carrots will also like at least one other vegetable is $\frac{24}{51} \approx 0.471$. The number of students who like cauliflower and broccoli is $4 + 3 = 7$. The number of students who like all three vegetables is 3. So, the probability that a student who likes cauliflower and broccoli will also like carrots is $\frac{3}{7} \approx 0.429$.

The number of students who do not like carrots, broccoli, or cauliflower is 23. The total number of students surveyed is 90. So, the probability that a student does not like any of the three vegetables is $23/90 \approx 0.256$.

137. C: Since each coin toss is an independent event, the probability of the compound event of flipping the coin three times is equal to the product of the probabilities of the individual events. For example, $P(HHH) = P(H) \cdot P(H) \cdot P(H)$, $P(HHT) = P(H) \cdot P(H) \cdot P(T)$, etc. When a coin is flipped three times, all of the possible outcomes are HHH, HHT, HTH, HTT, THH, THT, TTH, and TTT. Since the only way to obtain three heads is by the coin landing on heads three times,
$$P(three\ heads) = P(HHH) = P(H)P(H)P(H).$$
Likewise,
$$P(no\ heads) = P(T)P(T)P(T).$$
Since there are three ways to get one head,
$$P(one\ head) = P(HTT) + P(THT) + P(TTH) = P(H)P(T)P(T) + P(T)P(H)P(T) + P(T)P(T)P(H)$$
$$= P(H)[(3P(T)^2],$$
And since there are three ways to get two heads,
$$P(two\ heads) = P(HHT) + P(HTH) + P(THH)$$
$$= P(H)P(H)P(T) + P(H)P(T)P(H) + P(T)P(H)P(H) = P(H)^2[3P(T)]$$

Use these properties to calculate the experimental probability P(H):

30 out of 100 coin tosses resulted in three heads, and $P(three\ heads) = P(H)P(H)P(H) = P(H)^3$. So, experimental P(H) can be calculated by taking the cube root of $\frac{30}{100}$. $\sqrt[3]{0.3} \approx 0.67$. Similarly, $P(no\ heads) = P(T)P(T)P(T) = \frac{4}{100}$. $P(T) = \sqrt[3]{0.04} \approx 0.34$. $P(H) + P(T) = 1$, $P(T) = 1 - P(H)$. Thus, $P(H) = 1 - P(T) \approx 0.66$.

Notice that these calculated values of P(H) re approximately the same, Since 100 is a fairly large sample size for this kind of experiment, the approximation for $P(H)$ ought to consistent for the compiled data set. Rather than calculating $P(H)$ using the data for one head and two heads, use the average calculated probability to confirm that the number of expected outcomes of one head and two head matches the number of actual outcomes.

The number of expected outcomes of getting one head in three coin flips out of 100 trials $100\{0.665[3(1 - 0.665)^2]\} \approx 22$, and the expected outcome getting of two heads in three coin flips out of 100 trials three flips is $100\{0.665^2[3(1 - 0.665)]\} \approx 44$. Since 22 and 44 are, in fact, the data obtained, 0.665 is indeed a good approximation for P(H) when the coin used in this experiment is tossed.

138. D: A fair coin has a symmetrical binomial distribution which peaks in its center. Since choice B shows a skewed distribution for the fair coin, it cannot be the correct answer. From the frequency histogram given for the misshapen coin, it is evident that the misshapen coin is more likely to land on heads. Therefore, it is more likely that ten coin flips would result in fewer tails than ten coin flips of a fair coin; consequently, the probability distribution for the misshapen coin would be higher than the fair coin's distribution towards the left of the graph since the misshapen coin is less likely to land on tails. Choice A shows a probability distribution which peaks at a value of 5 and which is symmetrical with respect to the peak, which verifies that it cannot be correct. (Furthermore, in choice A, the sum of the probabilities shown for each number of tails for the misshapen coin is not equal to 1.) The distribution for the misshapen coin in choice C is skewed in the wrong direction, favoring tails instead of heads, and must therefore also be incorrect. Choice D shows the correct binomial distribution for the fair coin and the appropriate shift for the misshapen coin.

Another way to approach this question is to use the answer from the previous problem to determine the probability of obtaining particular events, such as no tails and no heads, and then compare those probabilities to the graphs. For example, for the misshapen coin, P(0 tails)=P(10 heads) $\approx (0.67)^{10}$, or 0.018, and the P(10 tails) $\approx (0.33)^{10}$, which is 0.000015. For a fair coin, P(0 tails)=$(0.5)^{10}$=P(0 heads). To find values other than these, it is helpful to use the binomial distribution formula $(_nC_r)p^r q^{n-r}$, where n is the number of trials, r is the number of successes, p is the probability of success, and q is the probability of failure. For this problem, obtaining tails is a success, and the probability of obtaining tails is $p = 0.33$ for the misshapen coin and $p = 0.5$ for the fair coin; so, $q = 0.67$ for the misshapen coin and $q = 0.5$ for the fair coin. To find the probability of, say, getting three tails for ten flips of the misshapen coin, find $(_nC_r)p^r q^{n-r} = (_{10}C_3)(0.33)^3(0.67)^7 = \frac{10!}{3!7!}(0.33)^3(0.67)^7 \approx 0.261$. The calculated probabilities match those shown in choice C.

139. C: When rolling two dice, there is only one way to roll a sum of two (rolling a 1 on each die) and twelve (rolling 6 on each die). In contrast, there are two ways to obtain a sum of three (rolling a 2 and 1 or a 1 and 2) and eleven (rolling a 5 and 6 or a 6 and 5), three ways to obtain a sum of four (1 and 3; 2 and 2; 3 and 1) or ten (4 and 6; 5 and 5; 6 and 4), and so on. Since the probability of obtaining each sum is inconsistent, choice C is not an appropriate simulation. Choice A is acceptable

since the probability of picking A, 1, 2, 3, 4, 5, 6, 7, 8, 9, or J from the modified deck cards of cards is equally likely, each with a probability of $\frac{4}{52-8} = \frac{4}{44} = \frac{1}{11}$. Choice B is also acceptable since the computer randomly generates one number from eleven possible numbers, so the probability of generating any of the numbers is $\frac{1}{11}$.

140. C: The number 00 represents the genotype aa. The numbers 11, 12, 21, and 22 represent the genotype bb.

```
28 93 97 37 92 00 27 21 87 13 62 63 10 31 55 09 47 07 54 88 38 88 10
98 34 01 45 14 34 46 38 61 93 22 37 39 57 03 93 50 53 16 28 65 81 60
21 12 13 10 19 91 04 18 49 01 99 30 11 16 00 48 04 63 59 24 02 42 23
06 32 52 19 18 94 94 46 63 87 41 79 39 85 20 43 20 15 03 39 33 77 45
66 77 70 92 25 27 68 71 89 35 98 55 85 47 60 97 12 92 53 44 45 41 51
22 09 23 81 33 04 35 43 48 32 80 36 95 64 56 34 74 55 37 64 84 51 50
25 99 51 94 19 46 10 44 17 25 75 52 47 35 70 65 08 50 98 09 02 24 30
59 00 03 21 40 30 86 16 53 91 28 17 97 58 75 76 73 83 54 40 54 13 38
36 67 74 80 63 12 41 27 96 61 66 05 60 69 96 15 56 82 57 31 83 26 24
78 42 76 49 56 06 57 78 67 02 96 40 82 29 14 07 29 62 90 31 08 26 71
61 18 22 84 23 33 49 29 90 07 08 05 14 59 72 86 44 69 68 99 06 11 95
43 72 58 28 93 97 37 92 00 27 21 87 13 62 61 15 31 55 09 47 07 54 88
38 88 10 98 34 01 45 14 34 46 38 61 93 22 37 39 57 03 93 50 53 16 28
65 81 60 21 12 13 10 19 91 04 18 49 01 99 30 11 16 00 48 04 63 59 24
02 42 23 06 32 52 19 18 94 94 46 63 87 41 79 39 85 20 43 20 15 03 39
33 77 45 66 77 70 92 25 27 68 71 89 35 98 55 85 47 60 97 12 92 53 44
45 41 51 22 09 23 81 33 04 35 43 48 32 80 36 95 64 56 34 74 55 37 64
84 51 50 25 99 51 94 19 46 10 44 17 25 75 52 47 35 70 65 08 50 98 09
02 24 30 59 00 03 21 40 30 86 16 53 91 28 17 97 58 75 76 73 83 54 40
54 13 38 36 67 74 80 63 12 41 27 96 61 66 05 60 69 96 15 56 82 57 31
83 26 24 78 42 76 49 56 06 57 78 67 02 96 40 82 29 14 07 29 62 90 31
08 26 71 61 18 22 84 23 33 49 29 90 07 08 05 14 59
```

There are six occurrences of 00, so the experimental probability of getting genotype aa is 6/500 = 0.012. There are 21 occurrences of 11, 12, 21, and 22, so the experimental probability of getting genotype bb is 21/500=0.042. The experimental probability of either getting genotype aa or bb is 0.012+0.042=0.054. Multiply this experimental probability by 100,000 to find the number of individuals expected to be homozygous for either allele in a population of 100,000. $0.054 \cdot 100,000 = 5,400$. Notice that this is higher than the expected number based on the theoretical probability. Since the allele frequencies are in a ratio of 1:2:7, the theoretical probability of getting either aa or bb is $\frac{1}{10} \cdot \frac{1}{10} + \frac{2}{10} \cdot \frac{2}{10} = \frac{5}{100} = 0.05$. Based on the theoretical probability, one would expect 5,000 members of a population of 100,000 to be homozygous for a or b.

141. D: A score of 85 is one standard deviation below the mean. Since approximately 68% of the data is within one standard deviation of the mean, about 32% (100%-68%) of the data is outside of one standard deviation within the mean. Normally distributed data is symmetric about the mean, which means that about 16% of the data lies below one standard deviation below the mean and about 16% of data lies above one standard deviation above the mean. Therefore, approximately 16% of individuals have IQs less than 85, while approximately 84% of the population has an IQ of at least 85. Since 84% of 300 is 252, about 252 people from the selected group have IQs of at least 85.

142. C: There are nine ways to assign the first digit since it can be any of the numbers 1-9. There are nine ways to assign the second digit since it can be any digit 0-9 EXCEPT for the digit assigned in place 1. There are eight ways to assign the third number since there are ten digits, two of which have already been assigned. There are seven ways to assign the fourth number, six ways to assign the fifth, five ways to assign the sixth, and four ways to assign the seventh. So, the number of combinations is $9 \cdot 9 \cdot 8 \cdot 7 \cdot 6 \cdot 5 \cdot 4 = 544{,}320$.

Another way to approach the problem is to notice that the arrangement of nine digits in the last six places is a sequence without reputation, or a permutation. (Note: this may be called a partial permutation since all of the elements of the set need not be used.) The number of possible sequences of a fixed length r of elements taken from a given set of size n is permutation $_nP_r = \frac{n!}{(n-r)!}$. So, the number of ways to arrange the last six digits is $_9P_6 = \frac{9!}{(9-6)!} = \frac{9!}{3!} = 60{,}480$. Multiply this number by nine since there are nine possibilities for the first digit of the phone number. $9 \cdot _9P_6 = 544{,}320$.

143. B: If each of the four groups in the class of twenty will contain three boys and two girls, there must be twelve boys and eight girls in the class. The number of ways the teacher can select three boys from a group of twelve boys is $_{12}C_3 = \frac{12!}{3!(12-3)!} = \frac{12!}{3!9!} = \frac{12 \cdot 11 \cdot 10 \cdot 9!}{3!9!} = \frac{12 \cdot 11 \cdot 10}{3 \cdot 2 \cdot 1} = 220$. The number of ways she can select two girls from a group of eight girls is $_8C_2 = \frac{8!}{2!(8-2)!} = \frac{8!}{2!6!} = \frac{8 \cdot 7 \cdot 6!}{2!6!} = \frac{8 \cdot 7}{2 \cdot 1} = 28$. Since each combination of boys can be paired with each combination of girls, the number of group combinations is $220 \cdot 28 = 6{,}160$.

144. B: One way to approach this problem is to first consider the number of arrangements of the five members of the family if Tasha (T) and Mac (M) must sit together. Treat them as a unit seated in a fixed location at the table; then arrange the other three family members (A, B, and C):

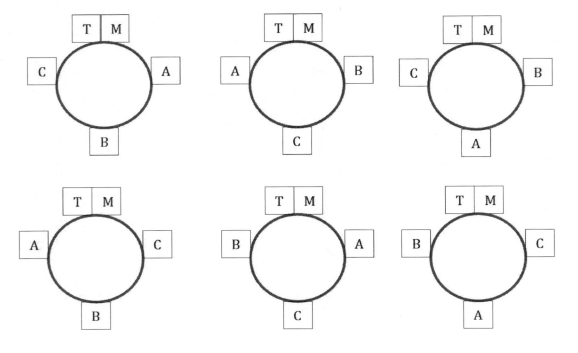

There are six ways to arrange four units around a circle as shown. (Any other arrangement would be a rotation in which the elements in the same order and would therefore not be a unique arrangement.) Note that there are $(n-1)!$ ways to arrange n units around a circle for $n > 1$.

Of course, Mac and Tasha are not actually a single unit. They would still be sitting beside each other if they were to trade seats, so there are twelve arrangements in which the two are seated next to one another. In all other arrangements of the five family members, they are separated. Therefore, to find the number of arrangements in which Tasha and Mac are not sitting together, subtract twelve from the possible arrangement of five units around a circle: $(5-1)! - 12 = 12$.

145. A: The recursive definition of the sequence gives the first term of the series, $a_1 = -1$. The definition also defines each term in the series as the sum of the previous term and 2. Therefore, the second term in the series is $-1 + 2 = 1$, the third term in the series is $1 + 2 = 3$, and so on.

n	a_n
1	-1
2	1
3	3

The relationship between n and a_n is linear, so the equation of the sequence can be found in the same way as the equation of a line. The value of a_n increases by two each time the value of n increases by 1.

n	$2n$	a_n
1	2	-1
2	4	1
3	6	3

Since the difference in $2n$ and a_n is 3, $a_n = 2n - 3$.

n	$2n - 3$	a_n
1	2-3	-1
2	4-3	1
3	6-3	3

146. B: The series is an infinite geometric series, the sum of which can be found by using the formula $\sum_{n=0}^{\infty} ar^n = \frac{a}{1-r}$, $|r| < 1$, where a is the first term in the series and r is the ratio between successive terms. In the series 200+100+50+25+ ..., $a = 200$ and $r = \frac{1}{2}$. So, the sum of the series is $\frac{200}{1-\frac{1}{2}} = \frac{200}{\frac{1}{2}} = 400$.

147. A: The sum of two vectors is equal to the sum of their components. Using component-wise addition, $v + w = (4 + (-3), 3 + 4) = (1,7)$. To multiply a vector by a scalar, multiply each component by that scalar. Using component-wise scalar multiplication, $2(1,7) = (2 \cdot 1, 2 \cdot 7) = (2,14)$.

148. A: First, subtract the two column matrices in parentheses by subtracting corresponding terms.

$$[2 \quad 0 \quad -5]\left(\begin{bmatrix} 4-3 \\ 2-5 \\ -1-(-5) \end{bmatrix}\right) = [2 \quad 0 \quad -5]\begin{bmatrix} 1 \\ -3 \\ 4 \end{bmatrix}$$

Then, multiply the matrices. The product of a 1×3 matrix and a 3×1 matrix is a 1×1 matrix.

$$[2 \quad 0 \quad -5]\begin{bmatrix} 1 \\ -3 \\ 4 \end{bmatrix} = [(2)(1) + (0)(-3) + (-5)(4)] = [-18]$$

Note that matrix multiplication is NOT commutative. The product of the 3x1 matrix $\begin{bmatrix} 1 \\ -3 \\ 4 \end{bmatrix}$ and the

1x3 matrix $[2 \quad 0 \quad -5]$ is the 3x3 matrix $\begin{bmatrix} 2 & 0 & -5 \\ -6 & 0 & 15 \\ 8 & 0 & -20 \end{bmatrix}$.

149. B: The table below shows the intersections of each set with each of the other sets.

Set	{2,4,6,8,10,12,...}	{1,2,3,4,6,12}	{1,2,4,9}
{2,4,6,8,10,12,... }	{2,4,6,8,10,12,...}	{2,4,6,12}	{2,4}
{1,2,3,4,6,12}	{2,4,6,12}	{1,2,3,4,6,12}	{1,2,4}
{1,2,4,9}	{2,4}	{1,2,4}	{1,2,4,9}

Notice that {2,4} is a subset of {2,4,6,12} and {1,2,4}. So, the intersection of {1,2,4,9} and the even integers is a subset of the intersection of the even integers and the factors of twelve, and the intersection of the set of even integers and {1,2,4,9} is a subset of the intersection of {1,2,4,9} and the factors of twelve. So, while it is not possible to determine which set is A and which is B, set C must be the set of factors of twelve: {1,2,3,4,6,12}.

150. D: Use a Venn diagram to help organize the given information. Start by filling in the space where the three circles intersect: Jenny tutored three students in all three areas. Use that information to fill in the spaces where two circles intersect: for example, she tutored four students in chemistry and for the ACT, and three of those were students she tutored in all three areas, so one student was tutored in chemistry and for the ACT but not for math. Once the diagram is completed, add the number of students who were tutored in all areas to the number of students tutored in only two of the three areas to the number of students tutored in only one area. The total number of students tutored was 3+2+2+1+3+2+1=14.

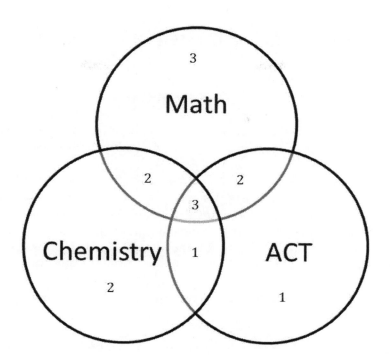

Secret Key #1 - Time is Your Greatest Enemy

Pace Yourself

Wear a watch. At the beginning of the test, check the time (or start a chronometer on your watch to count the minutes), and check the time after every few questions to make sure you are "on schedule."

If you are forced to speed up, do it efficiently. Usually one or more answer choices can be eliminated without too much difficulty. Above all, don't panic. Don't speed up and just begin guessing at random choices. By pacing yourself, and continually monitoring your progress against your watch, you will always know exactly how far ahead or behind you are with your available time. If you find that you are one minute behind on the test, don't skip one question without spending any time on it, just to catch back up. Take 15 fewer seconds on the next four questions, and after four questions you'll have caught back up. Once you catch back up, you can continue working each problem at your normal pace.

Furthermore, don't dwell on the problems that you were rushed on. If a problem was taking up too much time and you made a hurried guess, it must be difficult. The difficult questions are the ones you are most likely to miss anyway, so it isn't a big loss. It is better to end with more time than you need than to run out of time.

Lastly, sometimes it is beneficial to slow down if you are constantly getting ahead of time. You are always more likely to catch a careless mistake by working more slowly than quickly, and among very high-scoring test takers (those who are likely to have lots of time left over), careless errors affect the score more than mastery of material.

Secret Key #2 - Guessing is not Guesswork

You probably know that guessing is a good idea. Unlike other standardized tests, there is no penalty for getting a wrong answer. Even if you have no idea about a question, you still have a 20-25% chance of getting it right.

Most test takers do not understand the impact that proper guessing can have on their score. Unless you score extremely high, guessing will significantly contribute to your final score.

Monkeys Take the Test

What most test takers don't realize is that to insure that 20-25% chance, you have to guess randomly. If you put 20 monkeys in a room to take this test, assuming they answered once per question and behaved themselves, on average they would get 20-25% of the questions correct. Put 20 test takers in the room, and the average will be much lower among guessed questions. Why?

1. The test writers intentionally write deceptive answer choices that "look" right. A test taker has no idea about a question, so he picks the "best looking" answer, which is often wrong. The monkey has no idea what looks good and what doesn't, so it will consistently be right about 20-25% of the time.
2. Test takers will eliminate answer choices from the guessing pool based on a hunch or intuition. Simple but correct answers often get excluded, leaving a 0% chance of being correct. The monkey has no clue, and often gets lucky with the best choice.

This is why the process of elimination endorsed by most test courses is flawed and detrimental to your performance. Test takers don't guess; they make an ignorant stab in the dark that is usually worse than random.

$5 Challenge

Let me introduce one of the most valuable ideas of this course—the $5 challenge:

You only mark your "best guess" if you are willing to bet $5 on it.
You only eliminate choices from guessing if you are willing to bet $5 on it.

Why $5? Five dollars is an amount of money that is small yet not insignificant, and can really add up fast (20 questions could cost you $100). Likewise, each answer choice on one question of the test will have a small impact on your overall score, but it can really add up to a lot of points in the end.

The process of elimination IS valuable. The following shows your chance of guessing it right:

If you eliminate wrong answer choices until only this many remain:	Chance of getting it correct:
1	100%
2	50%
3	33%

However, if you accidentally eliminate the right answer or go on a hunch for an incorrect answer, your chances drop dramatically—to 0%. By guessing among all the answer choices, you are GUARANTEED to have a shot at the right answer.

That's why the $5 test is so valuable. If you give up the advantage and safety of a pure guess, it had better be worth the risk.

What we still haven't covered is how to be sure that whatever guess you make is truly random. Here's the easiest way:

Always pick the first answer choice among those remaining.

Such a technique means that you have decided, **before you see a single test question**, exactly how you are going to guess, and since the order of choices tells you nothing about which one is correct, this guessing technique is perfectly random.

This section is not meant to scare you away from making educated guesses or eliminating choices; you just need to define when a choice is worth eliminating. The $5 test, along with a pre-defined random guessing strategy, is the best way to make sure you reap all of the benefits of guessing.

Secret Key #3 - Practice Smarter, Not Harder

Many test takers delay the test preparation process because they dread the awful amounts of practice time they think necessary to succeed on the test. We have refined an effective method that will take you only a fraction of the time.

There are a number of "obstacles" in the path to success. Among these are answering questions, finishing in time, and mastering test-taking strategies. All must be executed on the day of the test at peak performance, or your score will suffer. The test is a mental marathon that has a large impact on your future.

Just like a marathon runner, it is important to work your way up to the full challenge. So first you just worry about questions, and then time, and finally strategy:

Success Strategy

1. Find a good source for practice tests.
2. If you are willing to make a larger time investment, consider using more than one study guide. Often the different approaches of multiple authors will help you "get" difficult concepts.
3. Take a practice test with no time constraints, with all study helps, "open book." Take your time with questions and focus on applying strategies.
4. Take a practice test with time constraints, with all guides, "open book."
5. Take a final practice test without open material and with time limits.

If you have time to take more practice tests, just repeat step 5. By gradually exposing yourself to the full rigors of the test environment, you will condition your mind to the stress of test day and maximize your success.

Secret Key #4 - Prepare, Don't Procrastinate

Let me state an obvious fact: if you take the test three times, you will probably get three different scores. This is due to the way you feel on test day, the level of preparedness you have, and the version of the test you see. Despite the test writers' claims to the contrary, some versions of the test WILL be easier for you than others.

Since your future depends so much on your score, you should maximize your chances of success. In order to maximize the likelihood of success, you've got to prepare in advance. This means taking practice tests and spending time learning the information and test taking strategies you will need to succeed.

Never go take the actual test as a "practice" test, expecting that you can just take it again if you need to. Take all the practice tests you can on your own, but when you go to take the official test, be prepared, be focused, and do your best the first time!

Secret Key #5 - Test Yourself

Everyone knows that time is money. There is no need to spend too much of your time or too little of your time preparing for the test. You should only spend as much of your precious time preparing as is necessary for you to get the score you need.

Once you have taken a practice test under real conditions of time constraints, then you will know if you are ready for the test or not.

If you have scored extremely high the first time that you take the practice test, then there is not much point in spending countless hours studying. You are already there.

Benchmark your abilities by retaking practice tests and seeing how much you have improved. Once you consistently score high enough to guarantee success, then you are ready.

If you have scored well below where you need, then knuckle down and begin studying in earnest. Check your improvement regularly through the use of practice tests under real conditions. Above all, don't worry, panic, or give up. The key is perseverance!

Then, when you go to take the test, remain confident and remember how well you did on the practice tests. If you can score high enough on a practice test, then you can do the same on the real thing.

General Strategies

The most important thing you can do is to ignore your fears and jump into the test immediately. Do not be overwhelmed by any strange-sounding terms. You have to jump into the test like jumping into a pool—all at once is the easiest way.

Make Predictions

As you read and understand the question, try to guess what the answer will be. Remember that several of the answer choices are wrong, and once you begin reading them, your mind will immediately become cluttered with answer choices designed to throw you off. Your mind is typically the most focused immediately after you have read the question and digested its contents. If you can, try to predict what the correct answer will be. You may be surprised at what you can predict.

Quickly scan the choices and see if your prediction is in the listed answer choices. If it is, then you can be quite confident that you have the right answer. It still won't hurt to check the other answer choices, but most of the time, you've got it!

Answer the Question

It may seem obvious to only pick answer choices that answer the question, but the test writers can create some excellent answer choices that are wrong. Don't pick an answer just because it sounds right, or you believe it to be true. It MUST answer the question. Once you've made your selection, always go back and check it against the question and make sure that you didn't misread the question and that the answer choice does answer the question posed.

Benchmark

After you read the first answer choice, decide if you think it sounds correct or not. If it doesn't, move on to the next answer choice. If it does, mentally mark that answer choice. This doesn't mean that you've definitely selected it as your answer choice, it just means that it's the best you've seen thus far. Go ahead and read the next choice. If the next choice is worse than the one you've already selected, keep going to the next answer choice. If the next choice is better than the choice you've already selected, mentally mark the new answer choice as your best guess.

The first answer choice that you select becomes your standard. Every other answer choice must be benchmarked against that standard. That choice is correct until proven otherwise by another answer choice beating it out. Once you've decided that no other answer choice seems as good, do one final check to ensure that your answer choice answers the question posed.

Valid Information

Don't discount any of the information provided in the question. Every piece of information may be necessary to determine the correct answer. None of the information in the question is there to throw you off (while the answer choices will certainly have information to throw you off). If two seemingly unrelated topics are discussed, don't ignore either. You can be confident there is a relationship, or it wouldn't be included in the question, and you are probably going to have to determine what is that relationship to find the answer.

Avoid "Fact Traps"

Don't get distracted by a choice that is factually true. Your search is for the answer that answers the question. Stay focused and don't fall for an answer that is true but irrelevant. Always go back to the question and make sure you're choosing an answer that actually answers the question and is not just a true statement. An answer can be factually correct, but it MUST answer the question asked. Additionally, two answers can both be seemingly correct, so be sure to read all of the answer choices, and make sure that you get the one that BEST answers the question.

Milk the Question

Some of the questions may throw you completely off. They might deal with a subject you have not been exposed to, or one that you haven't reviewed in years. While your lack of knowledge about the subject will be a hindrance, the question itself can give you many clues that will help you find the correct answer. Read the question carefully and look for clues. Watch particularly for adjectives and nouns describing difficult terms or words that you don't recognize. Regardless of whether you completely understand a word or not, replacing it with a synonym, either provided or one you more familiar with, may help you to understand what the questions are asking. Rather than wracking your mind about specific detailed information concerning a difficult term or word, try to use mental substitutes that are easier to understand.

The Trap of Familiarity

Don't just choose a word because you recognize it. On difficult questions, you may not recognize a number of words in the answer choices. The test writers don't put "make-believe" words on the test, so don't think that just because you only recognize all the words in one answer choice that that answer choice must be correct. If you only recognize words in one answer choice, then focus on that one. Is it correct? Try your best to determine if it is correct. If it is, that's great. If not, eliminate it. Each word and answer choice you eliminate increases your chances of getting the question correct, even if you then have to guess among the unfamiliar choices.

Eliminate Answers

Eliminate choices as soon as you realize they are wrong. But be careful! Make sure you consider all of the possible answer choices. Just because one appears right, doesn't mean that the next one won't be even better! The test writers will usually put more than one good answer choice for every question, so read all of them. Don't worry if you are stuck between two that seem right. By getting down to just two remaining possible choices, your odds are now 50/50. Rather than wasting too much time, play the odds. You are guessing, but guessing wisely because you've been able to knock out some of the answer choices that you know are wrong. If you are eliminating choices and realize that the last answer choice you are left with is also obviously wrong, don't panic. Start over and consider each choice again. There may easily be something that you missed the first time and will realize on the second pass.

Tough Questions

If you are stumped on a problem or it appears too hard or too difficult, don't waste time. Move on! Remember though, if you can quickly check for obviously incorrect answer choices, your chances of guessing correctly are greatly improved. Before you completely give up, at least try to knock out a couple of possible answers. Eliminate what you can and then guess at the remaining answer choices before moving on.

Brainstorm

If you get stuck on a difficult question, spend a few seconds quickly brainstorming. Run through the complete list of possible answer choices. Look at each choice and ask yourself, "Could this answer the question satisfactorily?" Go through each answer choice and consider it independently of the others. By systematically going through all possibilities, you may find something that you would otherwise overlook. Remember though that when you get stuck, it's important to try to keep moving.

Read Carefully

Understand the problem. Read the question and answer choices carefully. Don't miss the question because you misread the terms. You have plenty of time to read each question thoroughly and make sure you understand what is being asked. Yet a happy medium must be attained, so don't waste too much time. You must read carefully, but efficiently.

Face Value

When in doubt, use common sense. Always accept the situation in the problem at face value. Don't read too much into it. These problems will not require you to make huge leaps of logic. The test writers aren't trying to throw you off with a cheap trick. If you have to go beyond creativity and make a leap of logic in order to have an answer choice answer the question, then you should look at the other answer choices. Don't overcomplicate the problem by creating theoretical relationships or explanations that will warp time or space. These are normal problems rooted in reality. It's just that the applicable relationship or explanation may not be readily apparent and you have to figure things out. Use your common sense to interpret anything that isn't clear.

Prefixes

If you're having trouble with a word in the question or answer choices, try dissecting it. Take advantage of every clue that the word might include. Prefixes and suffixes can be a huge help. Usually they allow you to determine a basic meaning. Pre- means before, post- means after, pro - is positive, de- is negative. From these prefixes and suffixes, you can get an idea of the general meaning of the word and try to put it into context. Beware though of any traps. Just because con- is the opposite of pro-, doesn't necessarily mean congress is the opposite of progress!

Hedge Phrases

Watch out for critical hedge phrases, led off with words such as "likely," "may," "can," "sometimes," "often," "almost," "mostly," "usually," "generally," "rarely," and "sometimes." Question writers insert these hedge phrases to cover every possibility. Often an answer choice will be wrong simply because it leaves no room for exception. Unless the situation calls for them, avoid answer choices that have definitive words like "exactly," and "always."

Switchback Words

Stay alert for "switchbacks." These are the words and phrases frequently used to alert you to shifts in thought. The most common switchback word is "but." Others include "although," "however," "nevertheless," "on the other hand," "even though," "while," "in spite of," "despite," and "regardless of."

New Information

Correct answer choices will rarely have completely new information included. Answer choices typically are straightforward reflections of the material asked about and will directly relate to the question. If a new piece of information is included in an answer choice that doesn't even seem to relate to the topic being asked about, then that answer choice is likely incorrect. All of the information needed to answer the question is usually provided for you in the question. You should not have to make guesses that are unsupported or choose answer choices that require unknown information that cannot be reasoned from what is given.

Time Management

On technical questions, don't get lost on the technical terms. Don't spend too much time on any one question. If you don't know what a term means, then odds are you aren't going to get much further since you don't have a dictionary. You should be able to immediately recognize whether or not you know a term. If you don't, work with the other clues that you have—the other answer choices and terms provided—but don't waste too much time trying to figure out a difficult term that you don't know.

Contextual Clues

Look for contextual clues. An answer can be right but not the correct answer. The contextual clues will help you find the answer that is most right and is correct. Understand the context in which a phrase or statement is made. This will help you make important distinctions.

Don't Panic

Panicking will not answer any questions for you; therefore, it isn't helpful. When you first see the question, if your mind goes blank, take a deep breath. Force yourself to mechanically go through the steps of solving the problem using the strategies you've learned.

Pace Yourself

Don't get clock fever. It's easy to be overwhelmed when you're looking at a page full of questions, your mind is full of random thoughts and feeling confused, and the clock is ticking down faster than you would like. Calm down and maintain the pace that you have set for yourself. As long as you are on track by monitoring your pace, you are guaranteed to have enough time for yourself. When you get to the last few minutes of the test, it may seem like you won't have enough time left, but if you only have as many questions as you should have left at that point, then you're right on track!

Answer Selection

The best way to pick an answer choice is to eliminate all of those that are wrong, until only one is left and confirm that is the correct answer. Sometimes though, an answer choice may immediately look right. Be careful! Take a second to make sure that the other choices are not equally obvious. Don't make a hasty mistake. There are only two times that you should stop before checking other answers. First is when you are positive that the answer choice you have selected is correct. Second is when time is almost out and you have to make a quick guess!

Check Your Work

Since you will probably not know every term listed and the answer to every question, it is important that you get credit for the ones that you do know. Don't miss any questions through careless mistakes. If at all possible, try to take a second to look back over your answer selection and make sure you've selected the correct answer choice and haven't made a costly careless mistake (such as marking an answer choice that you didn't mean to mark). The time it takes for this quick double check should more than pay for itself in caught mistakes.

Beware of Directly Quoted Answers

Sometimes an answer choice will repeat word for word a portion of the question or reference section. However, beware of such exact duplication. It may be a trap! More than likely, the correct choice will paraphrase or summarize a point, rather than being exactly the same wording.

Slang

Scientific sounding answers are better than slang ones. An answer choice that begins "To compare the outcomes…" is much more likely to be correct than one that begins "Because some people insisted…"

Extreme Statements

Avoid wild answers that throw out highly controversial ideas that are proclaimed as established fact. An answer choice that states the "process should used in certain situations, if…" is much more likely to be correct than one that states the "process should be discontinued completely." The first is a calm rational statement and doesn't even make a definitive, uncompromising stance, using a hedge word "if" to provide wiggle room, whereas the second choice is a radical idea and far more extreme.

Answer Choice Families

When you have two or more answer choices that are direct opposites or parallels, one of them is usually the correct answer. For instance, if one answer choice states "x increases" and another answer choice states "x decreases" or "y increases," then those two or three answer choices are very similar in construction and fall into the same family of answer choices. A family of answer choices consists of two or three answer choices, very similar in construction, but often with directly opposite meanings. Usually the correct answer choice will be in that family of answer choices. The "odd man out" or answer choice that doesn't seem to fit the parallel construction of the other answer choices is more likely to be incorrect.

Additional Bonus Material

Due to our efforts to try to keep this book to a manageable length, we've created a link that will give you access to all of your additional bonus material.

Please visit http://www.mometrix.com/bonus948/priimath5161 to access the information.

Made in the USA
Lexington, KY
13 February 2017